Lecture Notes in Computer Science   5389

Commenced Publication in 1973
Founding and Former Series Editors:
Gerhard Goos, Juris Hartmanis, and Ja

Sándor P. Fekete (Ed.)

Algorithmic Aspects of Wireless Sensor Networks

Fourth International Workshop
ALGOSENSORS 2008
Reykjavik, Iceland, July 2008
Revised Selected Papers

 Springer

Volume Editor

Sándor P. Fekete
Department of Computer Science
Braunschweig University of Technology
38106 Braunschweig, Germany
E-mail: s.fekete@tu-bs.de

Library of Congress Control Number: Applied for

CR Subject Classification (1998): F.2, C.2, E.1, G.2

LNCS Sublibrary: SL 5 – Computer Communication Networks
and Telecommunications

ISSN   0302-9743
ISBN-10   3-540-92861-8 Springer Berlin Heidelberg New York
ISBN-13   978-3-540-92861-4 Springer Berlin Heidelberg New York

springer.com

© Springer-Verlag Berlin Heidelberg 2008
Printed in Germany

Typesetting: Camera-ready by author, data conversion by Scientific Publishing Services, Chennai, India
Printed on acid-free paper   SPIN: 12590049   06/3180   5 4 3 2 1 0

Preface

Wireless ad-hoc sensor networks are a very active research subject, as they have high potential to provide diverse services to numerous important applications, including remote monitoring and tracking in environmental applications and low-maintenance ambient intelligence in everyday life. The effective and efficient realization of such large-scale, complex ad-hoc networking environments requires intensive, coordinated technical research and development efforts, especially in power-aware, scalable, robust wireless distributed protocols, due to the unusual application requirements and the severe resource constraints of the sensor devices.

On the other hand, a solid foundational background seems necessary for sensor networks to achieve their full potential. It is a challenge for abstract modeling, algorithmic design and analysis to achieve provably efficient, scalable and fault-tolerant realizations of such huge, highly dynamic, complex, non-conventional networks. Features including the extremely large number of sensor devices in the network, the severe power, computing and memory limitations, their dense, random deployment and frequent failures pose new, interesting challenges of great practical impact for abstract modeling, algorithmic design, analysis and implementation.

This workshop aimed at bringing together research contributions related to diverse algorithmic and complexity-theoretic aspects of wireless sensor networks. This was the fourth event in the series. ALGOSENSORS 2004 was held in Turku, Finland, ALGOSENSORS 2006 was held in Venice, Italy, and ALGOSENSORS 2007 was held in Wrocław, Poland. Since its beginning, ALGOSENSORS has been collocated with ICALP. Previous proceedings have appeared in the Springer LNCS series: vol. 3121 (2004), vol. 4240 (2006), and vol. 4837 (2007).

ALGOSENSORS 2008 was part of ICALP 2008 and was held on July 12 2008 in Reykjavik, Iceland. After a careful review by the Program Committee, 11 out of 27 submissions were accepted; in addition, a keynote speech was given by Roger Wattenhofer. The Program Committee appreciates the help of 35 external referees, who provided additional expertise. We are also thankful for the help of the sponsors (EU-project "FRONTS" and coalesenses), who supported the organization of the meeting as well as a best-paper award.

October 2008   Sándor P. Fekete

Organization

Conference and Program Chair

Sándor P. Fekete   Braunschweig University of Technology, Germany

Program Committee

Michael Beigl	Braunschweig University of Technololgy, Germany
Michael Bender	Stony Brook University, USA
Ioannis Chatzigiannakis	University of Patras and CTI, Greece
Josep Diaz	Technical University of Catalonia, Spain
Shlomi Dolev	Ben-Gurion University, Israel
Alon Efrat	University of Arizona, USA
Michael Elkin	Ben Gurion University, Israel
Sándor P. Fekete	Braunschweig University of Technology, Germany (Chair)
Stefan Fischer	University of Lübeck, Germany
Stefan Funke	University of Greifswald, Germany
Jie Gao	Stony Brook University, USA
Magnús Halldórsson	Reykjavik University, Iceland
Riko Jacob	TU Munich, Germany
Alexander Kröller	Braunschweig University of Technology, Germany
Fabian Kuhn	ETH Zurich, Switzerland
Mirosław Kutyłowski	Wrocław University of Technology, Poland
Alberto Marchetti-Spaccamela	University of Rome "La Sapienza", Italy
Friedhelm Meyer auf der Heide	Universität Paderborn, Germany
Thomas Moscibroda	Microsoft Research, USA
David Peleg	Weizmann Institute, Israel
Dennis Pfisterer	University of Lübeck, Germany
Andrea Richa	Arizona State University, USA
Paolo Santi	CNR - Pisa, Italy
Christian Scheideler	TU Munich, Germany
Subhash Suri	University of California at Santa Barbara, USA
Dorothea Wagner	K.I.T, Karlsruhe, Germany
Roger Wattenhofer	ETH Zurich, Switzerland

Steering Committee

Josep Diaz	Technical University of Catalonia, Spain
Jan van Leeuwen	Utrecht University, The Netherlands
Sotiris Nikoletseas	University of Patras and CTI, Greece (Chair)
Jose Rolim	University of Geneva, Switzerland
Paul Spirakis	University of Patras and CTI, Greece

Additional Referees

Dror Aiger
Eitan Bachmat
Leonid Barenboim
Claudia Becker
Vincenzo Bonifaci
Carsten Buschmann
Jacek Cichon
Bastian Degener
Bernhard Fuchs
Joachim Gehweiler
Seth Gilbert
Horst Hellbrück
Tom Kamphans
Bastian Katz
Marcin Kik
Mirosław Korzeniowski
Ralf Klasing
Marina Kopeetsky

Sol Lederer
Nissan Lev-Tov
Peter Mahlmann
Steffen Mecke
Calvin Newport
Melih Onus
Raphael Eidenbenz
Laurence Pilard
Michal Ren
Rik Sarka
Christiane Schmidt
Barbara Schneider
Paul Spirakis
Elias Vicari
Axel Wegener
Dengpan Zhou
Xianjin Zhu

Sponsoring Institutions

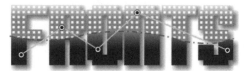

EU Commission: Project "FRONTS"
Contract Number: FP7 FET ICT-215270

coalesenses: Wireless Sensor Networks

Table of Contents

Algorithms for Sensor Networks:
What Is It Good for?

Roger Wattenhofer

Distributed Computing Group
Computer Engineering and Networks Laboratory
Information Technology and Electrical Engineering
ETH Zurich
Switzerland
wattenhofer@tik.ee.ethz.ch

Abstract. Absolutely nothing!? The merit of theory and algorithms in the context of wireless sensor and ad hoc networks is often questioned. Admittedly, coming up with theory success stories that will be accepted by practitioners is not easy. In my talk I will discuss the current score of the Theory vs. Practice game, after playing seven years for the Theory team. Probably due to a "seven year itch", I recently also started playing for the Practice team.

S. Fekete (Ed.): ALGOSENSORS 2008, LNCS 5389, p. 1, 2008.

Tight Local Approximation Results
for Max-Min Linear Programs

Patrik Floréen, Marja Hassinen, Petteri Kaski, and Jukka Suomela

Helsinki Institute for Information Technology HIIT
Helsinki University of Technology and University of Helsinki
P.O. Box 68, FI-00014 University of Helsinki, Finland
patrik.floreen@cs.helsinki.fi, marja.hassinen@cs.helsinki.fi,
petteri.kaski@cs.helsinki.fi, jukka.suomela@cs.helsinki.fi

Abstract. In a bipartite max-min LP, we are given a bipartite graph $\mathcal{G} = (V \cup I \cup K, E)$, where each agent $v \in V$ is adjacent to exactly one constraint $i \in I$ and exactly one objective $k \in K$. Each agent v controls a variable x_v. For each $i \in I$ we have a nonnegative linear constraint on the variables of adjacent agents. For each $k \in K$ we have a nonnegative linear objective function of the variables of adjacent agents. The task is to maximise the minimum of the objective functions. We study local algorithms where each agent v must choose x_v based on input within its constant-radius neighbourhood in $\mathcal{G}$. We show that for every $\epsilon > 0$ there exists a local algorithm achieving the approximation ratio $\Delta_I(1 - 1/\Delta_K) + \epsilon$. We also show that this result is the best possible – no local algorithm can achieve the approximation ratio $\Delta_I(1 - 1/\Delta_K)$. Here Δ_I is the maximum degree of a vertex $i \in I$, and Δ_K is the maximum degree of a vertex $k \in K$. As a methodological contribution, we introduce the technique of graph unfolding for the design of local approximation algorithms.

1   Introduction

As a motivating example, consider the task of data gathering in the following sensor network.

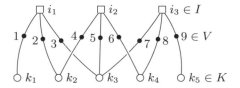

Each open circle is a sensor node $k \in K$, and each box is a relay node $i \in I$. The graph depicts the communication links between sensors and relays. Each sensor produces data which needs to be routed via adjacent relay nodes to a base station (not shown in the figure).

For each pair consisting of a sensor k and an adjacent relay i, we need to decide how much data is routed from k via i to the base station. For each such

S. Fekete (Ed.): ALGOSENSORS 2008, LNCS 5389, pp. 2–17, 2008.

decision, we introduce an *agent* $v \in V$; these are shown as black dots in the figure. We arrive at a bipartite graph $\mathcal{G}$ where the set of vertices is $V \cup I \cup K$ and each edge joins an agent $v \in V$ to a node $j \in I \cup K$.

Associated with each agent $v \in V$ is a variable x_v. Each relay constitutes a bottleneck: the relay has a limited battery capacity, which sets a limit on the total amount of data that can be forwarded through it. The task is to maximise the minimum amount of data gathered from a sensor node. In our example, the variable x_2 is the amount of data routed from the sensor k_2 via the relay i_1, the battery capacity of the relay i_1 is an upper bound for $x_1 + x_2 + x_3$, and the amount of data gathered from the sensor node k_2 is $x_2 + x_4$. Assuming that the maximum capacity of a relay is 1, the optimisation problem is to

$$
\begin{aligned}
\text{maximise} \quad & \min \{x_1, \ x_2 + x_4, \ x_3 + x_5 + x_7, \ x_6 + x_8, \ x_9\} \\
\text{subject to} \quad & x_1 + x_2 + x_3 \le 1, \\
& x_4 + x_5 + x_6 \le 1, \\
& x_7 + x_8 + x_9 \le 1, \\
& x_1, x_2, \ldots, x_9 \ge 0.
\end{aligned} \tag{1}
$$

In this work, we study *local algorithms* [1] for solving max-min linear programs (LPs) such as (1). In a local algorithm, each agent $v \in V$ must choose the value x_v solely based on its constant-radius neighbourhood in the graph $\mathcal{G}$. Such algorithms provide an extreme form of scalability in distributed systems; among others, a change in the topology of $\mathcal{G}$ affects the values x_v only in a constant-radius neighbourhood.

1.1 Max-Min Linear Programs

Let $\mathcal{G} = (V \cup I \cup K, E)$ be a bipartite, undirected communication graph where each edge $e \in E$ is of the form $\{v, j\}$ with $v \in V$ and $j \in I \cup K$. The elements $v \in V$ are called *agents*, the elements $i \in I$ are called *constraints*, and the elements $k \in K$ are called *objectives*; the sets V, I, and K are disjoint. We define $V_i = \{v \in V : \{v, i\} \in E\}$, $V_k = \{v \in V : \{v, k\} \in E\}$, $I_v = \{i \in I : \{v, i\} \in E\}$, and $K_v = \{k \in K : \{v, k\} \in E\}$ for all $i \in I$, $k \in K$, $v \in V$.

We assume that $\mathcal{G}$ is a bounded-degree graph; in particular, we assume that $|V_i| \le \Delta_I$ and $|V_k| \le \Delta_K$ for all $i \in I$ and $k \in K$ for some constants Δ_I and Δ_K.

A *max-min linear program* associated with $\mathcal{G}$ is defined as follows. Associate a variable x_v with each agent $v \in V$, associate a coefficient $a_{iv} \ge 0$ with each edge $\{i, v\} \in E$, $i \in I$, $v \in V$, and associate a coefficient $c_{kv} \ge 0$ with each edge $\{k, v\} \in E$, $k \in K$, $v \in V$. The task is to

$$
\begin{aligned}
\text{maximise} \quad & \omega = \min_{k \in K} \sum_{v \in V_k} c_{kv} x_v \\
\text{subject to} \quad & \sum_{v \in V_i} a_{iv} x_v \le 1 \qquad \forall i \in I, \\
& x_v \ge 0 \qquad \forall v \in V.
\end{aligned} \tag{2}
$$

We write ω^* for the optimum of (2).

1.2   Special Cases of Max-Min LPs

A max-min LP is a generalisation of a *packing LP*. Namely, in a packing LP there is only one linear nonnegative function to maximise, while in a max-min LP the goal is to maximise the minimum of multiple nonnegative linear functions.

Our main focus is on the *bipartite version* of the max-min LP problem. In the bipartite version we have $|I_v| = |K_v| = 1$ for each $v \in V$. We also define the 0/1 *version* [2]. In that case we have $a_{iv} = 1$ and $c_{kv} = 1$ for all $v \in V, i \in I_v, k \in K_v$. Our example (1) is both a bipartite max-min LP and a 0/1 max-min LP.

The *distance* between a pair of vertices $s, t \in V \cup I \cup K$ in $\mathcal{G}$ is the number of edges on a shortest path connecting s and t in $\mathcal{G}$. We write $B_{\mathcal{G}}(s, r)$ for the set of vertices within distance at most r from s. We say that $\mathcal{G}$ has *bounded relative growth* $1 + \delta$ *beyond radius* $R \in \mathbb{N}$ if

$$\frac{|V \cap B_{\mathcal{G}}(v, r+2)|}{|V \cap B_{\mathcal{G}}(v, r)|} \leq 1 + \delta \qquad \text{for all } v \in V, r \geq R.$$

Any bounded-degree graph $\mathcal{G}$ has a constant upper bound for δ. Regular grids are a simple example of a family of graphs where δ approaches 0 as R increases [3].

1.3   Local Algorithms and the Model of Computation

A local algorithm [1] is a distributed algorithm in which the output of a node is a function of input available within a fixed-radius neighbourhood; put otherwise, the algorithm runs in a constant number of communication rounds. In the context of distributed max-min LPs, the exact definition is as follows.

We say that the *local input* of a node $v \in V$ consists of the sets I_v and K_v and the coefficients a_{iv}, c_{kv} for all $i \in I_v, k \in K_v$. The local input of a node $i \in I$ consists of V_i and the local input of a node $k \in K$ consists of V_k. Furthermore, we assume that either (a) each node has a *unique identifier* given as part of the local input to the node [1,4]; or, (b) each vertex independently introduces an ordering of the edges incident to it. The latter, strictly weaker, assumption is often called *port numbering* [5]; in essence, each edge $\{s, t\}$ in $\mathcal{G}$ has two natural numbers associated with it: the port number in s and the port number in t.

Let $\mathcal{A}$ be a deterministic distributed algorithm executed by each of the nodes of $\mathcal{G}$ that finds a feasible solution x to any max-min LP (2) given locally as input to the nodes. Let $r \in \mathbb{N}$ be a constant independent of the input. We say that $\mathcal{A}$ is a *local algorithm* with *local horizon* r if, for every agent $v \in V$, the output x_v is a function of the local input of the nodes in $B_{\mathcal{G}}(v, r)$. Furthermore, we say that $\mathcal{A}$ has the *approximation ratio* $\alpha \geq 1$ if $\sum_{v \in V_k} c_{kv} x_v \geq \omega^*/\alpha$ for all $k \in K$.

1.4   Contributions and Prior Work

The following local approximability result is the main contribution of this paper.

Theorem 1. *For any* $\Delta_I \geq 2$, $\Delta_K \geq 2$, *and* $\epsilon > 0$, *there exists a local approximation algorithm for the bipartite max-min LP problem with the approximation ratio* $\Delta_I (1 - 1/\Delta_K) + \epsilon$. *The algorithm assumes only port numbering.*

We also show that the positive result of Theorem 1 is tight. Namely, we prove a matching lower bound on local approximability, which holds even if we assume both 0/1 coefficients and unique node identifiers.

Theorem 2. *For any $\Delta_I \geq 2$ and $\Delta_K \geq 2$, there exists no local approximation algorithm for the max-min LP problem with the approximation ratio $\Delta_I(1 - 1/\Delta_K)$. This holds even in the case of a bipartite, 0/1 max-min LP and with unique node identifiers given as input.*

Considering Theorem 1 in light of Theorem 2, we find it somewhat surprising that unique node identifiers are not required to obtain the best possible local approximation algorithm for bipartite max-min LPs.

In terms of earlier work, Theorem 1 is an improvement on the *safe algorithm* [3,6] which achieves the approximation ratio Δ_I. Theorem 2 improves upon the earlier lower bound $(\Delta_I + 1)/2 - 1/(2\Delta_K - 2)$ [3]; here it should be noted that our definition of the local horizon differs by a constant factor from earlier work [3] due to the fact that we have adopted a more convenient graph representation instead of a hypergraph representation.

In the context of packing and covering LPs, it is known [7] that any approximation ratio $\alpha > 1$ can be achieved by a local algorithm, assuming a bounded-degree graph and bounded coefficients. Compared with this, the factor $\Delta_I(1 - 1/\Delta_K)$ approximation in Theorem 1 sounds somewhat discouraging considering practical applications. However, the constructions that we use in our negative results are arguably far from the structure of, say, a typical real-world wireless network. In prior work [3] we presented a local algorithm that achieves a factor $1 + (2 + o(1))\delta$ approximation assuming that $\mathcal{G}$ has bounded relative growth $1 + \delta$ beyond some constant radius R; for a small δ, this is considerably better than $\Delta_I(1 - 1/\Delta_K)$ for general graphs. We complement this line of research on bounded relative growth graphs with a negative result that matches the prior positive result [3] up to constants.

Theorem 3. *Let $\Delta_I \geq 3$, $\Delta_K \geq 3$, and $0 < \delta < 1/10$. There exists no local approximation algorithm for the max-min LP problem with an approximation ratio less than $1 + \delta/2$. This holds even in the case of a bipartite max-min LP where the graph $\mathcal{G}$ has bounded relative growth $1 + \delta$ beyond some constant radius R.*

From a technical perspective, the proof of Theorem 1 relies on two ideas: *graph unfolding* and the idea of *averaging local solutions* of local LPs.

We introduce the unfolding technique in Sect. 2. In essence, we expand the finite input graph $\mathcal{G}$ into a possibly infinite tree $\mathcal{T}$. Technically, $\mathcal{T}$ is the *universal covering* of $\mathcal{G}$ [5]. While such unfolding arguments have been traditionally used to obtain impossibility results [8] in the context of distributed algorithms, here we use such an argument to simplify the design of local algorithms. In retrospect, our earlier approximation algorithm for 0/1 max-min LPs [2] can be interpreted as an application of the unfolding technique.

The idea of averaging local LPs has been used commonly in prior work on distributed algorithms [3,7,9,10]. Our algorithm can also be interpreted as a generalisation of the safe algorithm [6] beyond local horizon $r = 1$.

To obtain our negative results – Theorems 2 and 3 – we use a construction based on regular high-girth graphs. Such graphs [11,12,13,14] have been used in prior work to obtain impossibility results related to local algorithms [4,7,15].

2   Graph Unfolding

Let $\mathcal{H} = (V, E)$ be a connected undirected graph and let $v \in V$. Construct a (possibly infinite) rooted tree $\mathcal{T}_v = (\bar{V}, \bar{E})$ and a labelling $f_v \colon \bar{V} \to V$ as follows. First, introduce a vertex $\bar{v}$ as the root of $\mathcal{T}_v$ and set $f_v(\bar{v}) = v$. Then, for each vertex u adjacent to v in $\mathcal{H}$, add a new vertex $\bar{u}$ as a child of $\bar{v}$ and set $f_v(\bar{u}) = u$. Then expand recursively as follows. For each unexpanded $\bar{t} \neq \bar{v}$ with parent $\bar{s}$, and each $u \neq f(\bar{s})$ adjacent to $f(\bar{t})$ in $\mathcal{H}$, add a new vertex $\bar{u}$ as a child of $\bar{t}$ and set $f_v(\bar{u}) = u$. Mark $\bar{t}$ as expanded.

This construction is illustrated in Fig. 1. Put simply, we traverse $\mathcal{H}$ in a breadth-first manner and treat vertices revisited due to a cycle as new vertices; in particular, the tree $\mathcal{T}_v$ is finite if and only if $\mathcal{H}$ is acyclic.

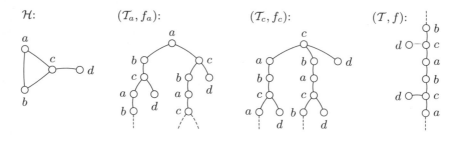

Fig. 1. An example graph $\mathcal{H}$ and its unfolding $(\mathcal{T}, f)$

The rooted, labelled trees $(\mathcal{T}_v, f_v)$ obtained in this way for different choices of $v \in V$ are isomorphic viewed as unrooted trees [5]. For example, the infinite labelled trees $(\mathcal{T}_a, f_a)$ and $(\mathcal{T}_c, f_c)$ in Fig. 1 are isomorphic and can be transformed into each other by rotations. Thus, we can define the *unfolding* of $\mathcal{H}$ as the labelled tree $(\mathcal{T}, f)$ where $\mathcal{T}$ is the unrooted version of $\mathcal{T}_v$ and $f = f_v$; up to isomorphism, this is independent of the choice of $v \in V$.

2.1   Unfolding in Graph Theory and Topology

We briefly summarise the graph theoretic and topological background related to the unfolding $(\mathcal{T}, f)$ of $\mathcal{H}$.

From a graph theoretic perspective, using the terminology of Godsil and Royle [17, §6.8], the surjection f is a homomorphism from $\mathcal{T}$ to $\mathcal{H}$. Moreover, it is a *local isomorphism*: the neighbours of $\bar{v} \in \bar{V}$ are in one-to-one correspondence with the neighbours of $f(\bar{v}) \in V$. A surjective local isomorphism f is a *covering map* and $(\mathcal{T}, f)$ is a *covering graph* of $\mathcal{H}$.

Covering maps in graph theory can be interpreted as a special case of covering maps in topology: $\mathcal{T}$ is a *covering space* of $\mathcal{H}$ and f is, again, a covering map. See, e.g., Hocking and Young [18, §4.8] or Munkres [19, §53].

In topology, a simply connected covering space is called a *universal covering space* [18, §4.8], [19, §80]. An analogous graph-theoretic concept is a tree: unfolding $\mathcal{T}$ of $\mathcal{H}$ is equal to the *universal covering* $\mathcal{U}(\mathcal{H})$ of $\mathcal{H}$ as defined by Angluin [5].

Unfortunately, the term "covering" is likely to cause confusion in the context of graphs. The term "lift" has been used for a covering graph [13,20]. We have borrowed the term "unfolding" from the field of model checking; see, e.g., Esparza and Heljanko [21].

2.2   Unfolding and Local Algorithms

Let us now view the graph $\mathcal{H}$ as the communication graph of a distributed system, and let $(\mathcal{T}, f)$ be the unfolding of $\mathcal{H}$. Even if $\mathcal{T}$ in general is countably infinite, a local algorithm $\mathcal{A}$ with local horizon r can be designed to operate at a node of $v \in \mathcal{H}$ exactly *as if* it was a node $\bar{v} \in f^{-1}(v)$ in the communication graph $\mathcal{T}$. Indeed, assume that the local input at $\bar{v}$ is identical to the local input at $f(\bar{v})$, and observe that the radius r neighbourhood of the node $\bar{v}$ in $\mathcal{T}$ is equal to the rooted tree $\mathcal{T}_v$ trimmed to depth r; let us denote this by $\mathcal{T}_v(r)$. To gather the information in $\mathcal{T}_v(r)$, it is sufficient to gather information on all walks of length at most r starting at v in $\mathcal{H}$; using port numbering, the agents can detect and discard walks that consecutively traverse the same edge.

Assuming that only port numbering is available, the information in $\mathcal{T}_v(r)$ is in fact *all* that the agent v can gather. Indeed, to assemble, say, the subgraph of $\mathcal{H}$ induced by $B_{\mathcal{H}}(v, r)$, the agent v in general needs to distinguish between a short cycle and a long path, and these are indistinguishable without node identifiers.

2.3   Unfolding and Max-Min LPs

Let us now consider a max-min LP associated with a graph $\mathcal{G}$. The unfolding of $\mathcal{G}$ leads in a natural way to the unfolding of the max-min LP. We show in this section that in order to prove Theorem 1, it is sufficient to design a local approximation algorithm for unfoldings of a max-min LP.

Unfolding requires us to consider max-min LPs where the underlying communication graph is countably infinite. The graph is always a bounded-degree graph, however. This allows us to circumvent essentially all of the technicalities otherwise encountered with infinite problem instances; cf. Anderson and Nash [16]. For the purposes of this work, it suffices to define that x is a *feasible solution with utility at least* ω if (x, ω) satisfies

$$\begin{aligned}
\sum_{v \in V_k} c_{kv} x_v &\geq \omega & \forall k \in K, \\
\sum_{v \in V_i} a_{iv} x_v &\leq 1 & \forall i \in I, \\
x_v &\geq 0 & \forall v \in V.
\end{aligned} \tag{3}$$

Observe that each of the sums in (3) is finite. Furthermore, this definition is compatible with the finite max-min LP defined in Sect. 1.1. Namely, if ω^* is the optimum of a finite max-min LP, then there exists a feasible solution x^* with utility at least ω^*.

Let $\mathcal{G} = (V \cup I \cup K, E)$ be the underlying finite communication graph. Unfold $\mathcal{G}$ to obtain a (possibly infinite) tree $\mathcal{T} = (\bar{V} \cup \bar{I} \cup \bar{K}, \bar{E})$ with a labelling f. Extend this to an unfolding of the max-min LP by associating a variable $x_{\bar{v}}$ with each agent $\bar{v} \in \bar{V}$, the coefficient $a_{\bar{\iota}\bar{v}} = a_{f(\bar{\iota}),f(\bar{v})}$ for each edge $\{\bar{\iota}, \bar{v}\} \in \bar{E}$, $\bar{\iota} \in \bar{I}, \bar{v} \in \bar{V}$, and the coefficient $c_{\bar{\kappa}\bar{v}} = c_{f(\bar{\kappa}),f(\bar{v})}$ for each edge $\{\bar{\kappa}, \bar{v}\} \in \bar{E}, \bar{\kappa} \in \bar{K}$, $\bar{v} \in \bar{V}$. Furthermore, assume an arbitrary port numbering for the edges incident to each of the nodes in $\mathcal{G}$, and extend this to a port numbering for the edges incident to each of the nodes in $\mathcal{T}$ so that the port numbers at the ends of each edge $\{\bar{u}, \bar{v}\} \in \bar{E}$ are identical to the port numbers at the ends of $\{f(\bar{u}), f(\bar{v})\}$.

Lemma 1. *Let $\bar{A}$ be a local algorithm for unfoldings of a family of max-min LPs and let $\alpha \geq 1$. Assume that the output x of $\bar{A}$ satisfies $\sum_{v \in V_k} c_{kv} x_v \geq \omega'/\alpha$ for all $k \in K$ if there exists a feasible solution with utility at least ω'. Furthermore, assume that $\bar{A}$ uses port numbering only. Then, there exists a local approximation algorithm A with the approximation ratio α for this family of max-min LPs.*

Proof. Let x^* be an optimal solution of the original instance, with utility ω^*. Set $x_{\bar{v}} = x^*_{f(\bar{v})}$ to obtain a solution of the unfolding. This is a feasible solution because the variables of the agents adjacent to a constraint $\bar{\iota}$ in the unfolding have the same values as the variables of the agents adjacent to the constraint $f(\bar{\iota})$ in the original instance. By similar reasoning, we can show that this is a feasible solution with utility at least ω^*.

Construct the local algorithm A using the assumed algorithm $\bar{A}$ as follows. Each node $v \in V$ simply behaves as if it was a node $\bar{v} \in f^{-1}(v)$ in the unfolding $\mathcal{T}$ and simulates $\bar{A}$ for $\bar{v}$ in $\mathcal{T}$. By assumption, the solution x computed by $\bar{A}$ in the unfolding has to satisfy

$$\sum_{\bar{v} \in V_{\bar{\kappa}}} c_{\bar{\kappa}\bar{v}} x_{\bar{v}} \geq \omega^*/\alpha \qquad \forall \bar{\kappa} \in \bar{K},$$
$$\sum_{\bar{v} \in V_{\bar{\iota}}} a_{\bar{\iota}\bar{v}} x_{\bar{v}} \leq 1 \qquad \forall \bar{\iota} \in \bar{I}.$$

Furthermore, if $f(\bar{u}) = f(\bar{v})$ for $\bar{u}, \bar{v} \in \bar{V}$, then the neighbourhoods of $\bar{u}$ and $\bar{v}$ contain precisely the same information (including the port numbering), so the deterministic $\bar{A}$ must output the same value $x_{\bar{u}} = x_{\bar{v}}$. Giving the output $x_v = x_{\bar{v}}$ for any $\bar{v} \in f^{-1}(v)$ therefore yields a feasible, α-approximate solution to the original instance. ∎

We observe that Lemma 1 generalises beyond max-min LPs; we did not exploit the linearity of the constraints and the objectives.

3 Approximability Results

We proceed to prove Theorem 1. Let $\Delta_I \geq 2$, $\Delta_K \geq 2$, and $\epsilon > 0$ be fixed. By virtue of Lemma 1, it suffices to consider only bipartite max-min LPs where the graph $\mathcal{G}$ is a (finite or countably infinite) tree.

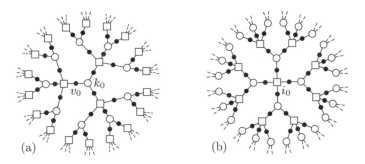

Fig. 2. Radius 6 neighbourhoods of (a) an objective $k_0 \in K$ and (b) a constraint $i_0 \in I$ in the regularised tree $\mathcal{G}$, assuming $\Delta_I = 4$ and $\Delta_K = 3$. The black dots represent agents $v \in V$, the open circles represent objectives $k \in K$, and the boxes represent constraints $i \in I$.

To ease the analysis, it will be convenient to *regularise* $\mathcal{G}$ to a countably infinite tree with $|V_i| = \Delta_I$ and $|V_k| = \Delta_K$ for all $i \in I$ and $k \in K$.

To this end, if $|V_i| < \Delta_I$ for some $i \in I$, add $\Delta_I - |V_i|$ new *virtual* agents as neighbours of i. Let v be one of these agents. Set $a_{iv} = 0$ so that no matter what value one assigns to x_v, it does not affect the feasibility of the constraint i. Then add a new virtual objective k adjacent to v and set, for example, $c_{kv} = 1$. As one can assign an arbitrarily large value to x_v, the virtual objective k will not be a bottleneck.

Similarly, if $|V_k| < \Delta_K$ for some $k \in K$, add $\Delta_K - |V_k|$ new virtual agents as neighbours of k. Let v be one of these agents. Set $c_{kv} = 0$ so that no matter what value one assigns to x_v, it does not affect the value of the objective k. Then add a new virtual constraint i adjacent to v and set, for example, $a_{iv} = 1$.

Now repeat these steps and grow virtual trees rooted at the constraints and objectives that had less than Δ_I or Δ_K neighbours. The result is a countably infinite tree where $|V_i| = \Delta_I$ and $|V_k| = \Delta_K$ for all $i \in I$ and $k \in K$. Observe also that from the perspective of a local algorithm it suffices to grow the virtual trees only up to depth r because then the radius r neighbourhood of each original node is indistinguishable from the regularised tree. The resulting topology is illustrated in Fig. 2 from the perspective of an original objective $k_0 \in K$ and an original constraint $i_0 \in I$.

3.1 Properties of Regularised Trees

For each $v \in V$ in a regularised tree $\mathcal{G}$, define $K(v, \ell) = K \cap B_{\mathcal{G}}(v, 4\ell+1)$, that is, the set of objectives k within distance $4\ell+1$ from v. For example, $K(v, 1)$ consists of 1 objective at distance 1, $\Delta_I - 1$ objectives at distance 3, and $(\Delta_K - 1)(\Delta_I - 1)$ objectives at distance 5; see Fig. 2a. In general, we have

$$|K(v, \ell)| = 1 + (\Delta_I - 1)\Delta_K n(\ell), \tag{4}$$

where

$$n(\ell) = \sum_{j=0}^{\ell-1} (\Delta_I - 1)^j (\Delta_K - 1)^j.$$

Let $k \in K$. If $u, v \in V_k$, $u \neq v$, then the objective at distance 1 from u is the same as the objective at distance 1 from v; therefore $K(u, 0) = K(v, 0)$. The objectives at distance 3 from u are at distance 5 from v, and the objectives at distance 5 from u are at distance 3 or 5 from v; therefore $K(u, 1) - K(v, 1)$. By a similar reasoning, we obtain

$$K(u, \ell) = K(v, \ell) \qquad \forall \, \ell \in \mathbb{N}, \; k \in K, \; u, v \in V_k. \tag{5}$$

Let us then study a constraint $i \in I$. Define

$$K(i, \ell) = \bigcap_{v \in V_i} K(v, \ell) = K \cap B_{\mathcal{G}}(i, 4\ell) = K \cap B_{\mathcal{G}}(i, 4\ell - 2).$$

For example, $K(i, 2)$ consists of Δ_I objectives at distance 2 from the constraint i, and $\Delta_I(\Delta_K - 1)(\Delta_I - 1)$ objectives at distance 6 from the constraint i; see Fig. 2b. In general, we have

$$|K(i, \ell)| = \Delta_I n(\ell). \tag{6}$$

For adjacent $v \in V$ and $i \in I$, we also define $\partial K(v, i, \ell) = K(v, \ell) \setminus K(i, \ell)$. We have by (4) and (6)

$$|\partial K(v, i, \ell)| = 1 + (\Delta_I \Delta_K - \Delta_I - \Delta_K) \, n(\ell). \tag{7}$$

3.2   Local Approximation on Regularised Trees

It now suffices to meet Lemma 1 for bipartite max-min LPs in the case when the underlying graph $\mathcal{G}$ is a countably infinite regularised tree. To this end, let $L \in \mathbb{N}$ be a constant that we choose later; L depends only on Δ_I, Δ_K and ϵ.

Each agent $u \in V$ now executes the following algorithm. First, the agent gathers all objectives $k \in K$ within distance $4L + 1$, that is, the set $K(u, L)$. Then, for each $k \in K(u, L)$, the agent u gathers the radius $4L + 2$ neighbourhood of k; let $\mathcal{G}(k, L)$ be this subgraph. In total, the agent u accumulates information from distance $r = 8L + 3$ in the tree; this is the local horizon of the algorithm.

The structure of $\mathcal{G}(k, L)$ is a tree similar to the one shown in Fig. 2a. The leaf nodes of the tree $\mathcal{G}(k, L)$ are constraints. For each $k \in K(u, L)$, the agent u forms the constant-size *subproblem* of (2) restricted to the vertices of $\mathcal{G}(k, L)$ and solves it optimally using a deterministic algorithm; let x^{kL} be the solution. Once the agent u has solved the subproblem for every $k \in K(u, L)$, it sets

$$q = 1/\big(\Delta_I + \Delta_I(\Delta_I - 1)(\Delta_K - 1)n(L)\big), \tag{8}$$

$$x_u = q \sum_{k \in K(u, L)} x_u^{kL}. \tag{9}$$

This completes the description of the algorithm. In Sect. 3.3 we show that the computed solution x is feasible, and in Sect. 3.4 we establish a lower bound on the performance of the algorithm. Section 3.5 illustrates the algorithm with an example.

3.3   Feasibility

Because each x^{kL} is a feasible solution, we have

$$\sum_{v \in V_i} a_{iv} x_v^{kL} \leq 1 \qquad\qquad \forall \text{ non-leaf } i \in I \text{ in } \mathcal{G}(k,L), \qquad (10)$$

$$a_{iv} x_v^{kL} \leq 1 \qquad\qquad \forall \text{ leaf } i \in I, \ v \in V_i \text{ in } \mathcal{G}(k,L). \qquad (11)$$

Let $i \in I$. For each subproblem $\mathcal{G}(k,L)$ with $v \in V_i$, $k \in K(i,L)$, the constraint i is a non-leaf vertex; therefore

$$\sum_{v \in V_i} \sum_{k \in K(i,L)} a_{iv} x_v^{kL} = \sum_{k \in K(i,L)} \sum_{v \in V_i} a_{iv} x_v^{kL}$$

$$\overset{(10)}{\leq} \sum_{k \in K(i,L)} 1 \qquad\qquad (12)$$

$$\overset{(6)}{=} \Delta_I \, n(L).$$

For each subproblem $\mathcal{G}(k,L)$ with $v \in V_i$, $k \in \partial K(v,i,L)$, the constraint i is a leaf vertex; therefore

$$\sum_{v \in V_i} \sum_{k \in \partial K(v,i,L)} a_{iv} x_v^{kL} \overset{(11)}{\leq} \sum_{v \in V_i} \sum_{k \in \partial K(v,i,L)} 1 \qquad\qquad (13)$$

$$\overset{(7)}{=} \Delta_I \left(1 + (\Delta_I \Delta_K - \Delta_I - \Delta_K) \, n(L) \right).$$

Combining (12) and (13), we can show that the constraint i is satisfied:

$$\sum_{v \in V_i} a_{iv} x_v \overset{(9)}{=} q \sum_{v \in V_i} a_{iv} \sum_{k \in K(v,L)} x_v^{kL}$$

$$= q \left(\sum_{v \in V_i} \sum_{k \in K(i,L)} a_{iv} x_v^{kL} \right) + q \left(\sum_{v \in V_i} \sum_{k \in \partial K(v,i,L)} a_{iv} x_v^{kL} \right)$$

$$\leq q \Delta_I n(L) + q \Delta_I \left(1 + (\Delta_I \Delta_K - \Delta_I - \Delta_K) n(L) \right)$$

$$\overset{(8)}{=} 1.$$

3.4   Approximation Ratio

Consider an arbitrary feasible solution x' of the unrestricted problem (2) with utility at least ω'. This feasible solution is also a feasible solution of each finite subproblem restricted to $\mathcal{G}(k,L)$; therefore

$$\sum_{v \in V_h} c_{hv} x_v^{kL} \geq \omega' \qquad\qquad \forall \, h \in K \text{ in } \mathcal{G}(k,L). \qquad (14)$$

Define

$$\alpha = \frac{1}{q(1 + (\Delta_I - 1)\Delta_K n(L))}. \qquad (15)$$

Consider an arbitrary $k \in K$ and $u \in V_k$. We have

$$
\begin{aligned}
\sum_{v \in V_k} c_{kv} x_v &= q \sum_{v \in V_k} c_{kv} \sum_{h \in K(v,L)} x_v^{hL} \\
&\overset{(5)}{=} q \sum_{h \in K(u,L)} \sum_{v \in V_k} c_{kv} x_v^{hL} \\
&\overset{(14)}{\geq} q \sum_{h \in K(u,L)} \omega' \\
&\overset{(4)}{\geq} q(1 + (\Delta_I - 1)\Delta_K n(L)) \omega' \\
&\overset{(15)}{=} \omega'/\alpha.
\end{aligned}
$$

By (8) and (15), we have

$$
\alpha = \Delta_I \left(1 - \frac{1}{\Delta_K + 1/((\Delta_I - 1)n(L))} \right).
$$

For a sufficiently large L, we meet Lemma 1 with $\alpha < \Delta_I(1 - 1/\Delta_K) + \epsilon$. This completes the proof of Theorem 1.

3.5   An Example

Assume that $\Delta_I = 4$, $\Delta_K = 3$, and $L = 1$. For each $k \in K$, our approximation algorithm constructs and solves a subproblem; the structure of the subproblem is illustrated in Fig. 2a. Then we simply sum up the optimal solutions of each subproblem. For any $v \in V$, the variable x_v is involved in exactly $|K(v, L)| = 10$ subproblems.

First, consider an objective $k \in K$. The boundary of a subproblem always lies at a constraint, never at an objective. Therefore the objective k and all its adjacent agents $v \in V_k$ are involved in 10 subproblems. We satisfy the objective exactly 10 times, each time at least as well as in the global optimum.

Second, consider a constraint $i \in I$. The constraint may lie in the middle of a subproblem or at the boundary of a subproblem. The former happens in this case $|K(i, L)| = 4$ times; the latter happens $|V_i| \cdot |\partial K(v, i, L)| = 24$ times. In total, we use up the capacity available at the constraint i exactly 28 times. See Fig. 2b for an illustration; there are 28 objectives within distance 6 from the constraint $i_0 \in I$.

Finally, we scale down the solution by factor $q = 1/28$. This way we obtain a solution which is feasible and within factor $\alpha = 2.8$ of optimum. This is close to the lower bound $\alpha > 2.66$ from Theorem 2.

4   Inapproximability Results

We proceed to prove Theorems 2 and 3. Let $r = 4, 8, \ldots,$ $s \in \mathbb{N}$, $D_I \in \mathbb{Z}^+$, and $D_K \in \mathbb{Z}^+$ be constants whose values we choose later. Let $\mathcal{Q} = (I' \cup K', E')$ be a bipartite graph where the degree of each $i \in I'$ is D_I, the degree of each $k \in K'$ is D_K, and there is no cycle of length less than $g = 2(4s + 2 + r) + 1$. We first show that such graphs exist for all values of the parameters.

We say that a bipartite graph $\mathcal{G} = (V \cup U, E)$ is (a, b)-*regular* if the degree of each node in V is a and the degree of each node in U is b.

Lemma 2. *For any positive integers a, b and g, there exists an (a, b)-regular bipartite graph which has no cycle of length less than g.*

Proof (sketch). We slightly adapt a proof of a similar result for d-regular graphs [13, Theorem A.2] to our needs. We proceed by induction on g, for $g = 4, 6, 8, \ldots$.

For the base case $g = 4$, we can choose the complete bipartite graph $K_{b,a}$.

Next consider $g \geq 6$. Let $\mathcal{G} = (V \cup U, E)$ be an (a, b)-regular bipartite graph where the length of the shortest cycle is $c > g - 2$. Let $S \subseteq E$. Construct a graph $\mathcal{G}_S = (V_S \cup U_S, E_S)$ as follows:

$$V_S = \{0, 1\} \times V,$$
$$U_S = \{0, 1\} \times U,$$
$$E_S = \{\{(0, v), (0, u)\}, \{(1, v), (1, u)\} : \{v, u\} \in S\}$$
$$\cup \{\{(0, v), (1, u)\}, \{(1, v), (0, u)\} : \{v, u\} \in E \setminus S\}.$$

The graph $\mathcal{G}_S$ is an (a, b)-regular bipartite graph. Furthermore, $\mathcal{G}_S$ has no cycle of length less than c. We proceed to show that there exists a subset S such that the number of cycles of length exactly c in $\mathcal{G}_S$ is strictly less than the number of cycles of length c in $\mathcal{G}$. Then by a repeated application of the same construction, we can conclude that there exists a graph which is an (a, b)-regular bipartite graph and which has no cycle of length c; that is, its girth is at least g.

We use the probabilistic method to show that the number of cycles of length c decreases for some $S \subseteq E$. For each $e \in E$, toss an independent and unbiased coin to determine whether $e \in S$. For each cycle $C \subseteq E$ of length c in $\mathcal{G}$, we have in $\mathcal{G}_S$ either two cycles of length c or one cycle of length $2c$, depending on the parity of $|C \cap S|$. The expected number of cycles of length c in $\mathcal{G}_S$ is therefore equal to the number of cycles of length c in $\mathcal{G}$. The choice $S = E$ doubles the number of such cycles; therefore some other choice necessarily decreases the number of such cycles. □

4.1 The Instance $\mathcal{S}$

Given the graph $\mathcal{Q} = (I' \cup K', E')$, we construct an instance of the max-min LP problem, $\mathcal{S}$. The underlying communication graph $\mathcal{G} = (V \cup I \cup K, E)$ is constructed as shown in the following figure.

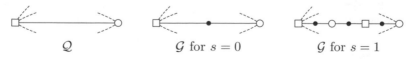

$\mathcal{Q}$ $\mathcal{G}$ for $s = 0$ $\mathcal{G}$ for $s = 1$

Each edge $e = \{i, k\} \in E'$ is replaced by a path of length $4s + 2$: the path begins with the constraint $i \in I'$; then there are s segments of agent–objective–agent–constraint; and finally there is an agent and the objective $k \in K'$. There are no other edges or vertices in $\mathcal{G}$. For example, in the case of $s = 0$, $D_I = 4$, $D_K = 3$, and sufficiently large g, the graph $\mathcal{G}$ looks *locally* similar to the trees in Fig. 2, even though there may be long cycles.

The coefficients of the instance $\mathcal{S}$ are chosen as follows. For each objective $k \in K'$, we set $c_{kv} = 1$ for all $v \in V_k$. For each objective $k \in K \setminus K'$, we set $c_{kv} = D_K - 1$ for all $v \in V_k$. For each constraint $i \in I$, we set $a_{iv} = 1$. Observe that $\mathcal{S}$ is a bipartite max-min LP; furthermore, in the case $s = 0$, this is a 0/1 max-min LP. We can choose the port numbering in $\mathcal{G}$ in an arbitrary manner, and we can assign unique node identifiers to the vertices of $\mathcal{G}$ as well.

Lemma 3. *The utility of any feasible solution of $\mathcal{S}$ is at most*

$$\frac{D_K}{D_I} \cdot \frac{D_K - 1 + D_K D_I s - D_I s}{D_K - 1 + D_K s}.$$

Proof. Consider a feasible solution x of $\mathcal{S}$, with utility ω. We proceed to derive an upper bound for ω. For each $j = 0, 1, \ldots, 2s$, let $V(j)$ consist of agents $v \in V$ such that the distance to the nearest constraint $i \in I'$ is $2j + 1$. That is, $V(0)$ consists of the agents adjacent to an $i \in I'$ and $V(2s)$ consists of the agents adjacent to a $k \in K'$. Let $m = |E'|$; we observe that $|V(j)| = m$ for each j.

Let $X(j) = \sum_{v \in V(j)} x_v / m$. From the constraints $i \in I'$ we obtain

$$X(0) = \sum_{v \in V(0)} x_v / m = \sum_{i \in I'} \sum_{v \in V_i} a_{iv} x_v / m \leq \sum_{i \in I'} 1/m = |I'|/m = 1/D_I.$$

Similarly, from the objectives $k \in K'$ we obtain $X(2s) \geq \omega |K'|/m = \omega/D_K$.

From the objectives $k \in K \setminus K'$, taking into account our choice of the coefficients c_{kv}, we obtain the inequality $X(2t) + X(2t + 1) \geq \omega/(D_K - 1)$ for $t = 0, 1, \ldots, s - 1$. From the constraints $i \in I \setminus I'$, we obtain the inequality $X(2t + 1) + X(2t + 2) \leq 1$ for $t = 0, 1, \ldots, s - 1$. Combining inequalities, we have

$$\omega/D_K - 1/D_I \leq X(2s) - X(0)$$
$$= \sum_{t=0}^{s-1} \Big(\big(X(2t+1) + X(2t+2) \big) - \big(X(2t) + X(2t+1) \big) \Big)$$
$$\leq s \cdot \big(1 - \omega/(D_K - 1) \big).$$

The claim follows. □

4.2 The Instance $\mathcal{S}_k$

Let $k \in K'$. We construct another instance of the max-min LP problem, $\mathcal{S}_k$. The communication graph of $\mathcal{S}_k$ is the subgraph $\mathcal{G}_k$ of $\mathcal{G}$ induced by $B_{\mathcal{G}}(k, 4s+2+r)$. By the choice of g, there is no cycle in $\mathcal{G}_k$. As r is a multiple of 4, the leaves of the tree $\mathcal{G}_k$ are constraints. For example, in the case of $s = 0$, $D_I = 4$, $D_K = 3$, and $r = 4$, the graph $\mathcal{G}_k$ is isomorphic to the tree of Fig. 2a. The coefficients, port numbers and node identifiers are chosen in $\mathcal{G}_k$ exactly as in $\mathcal{G}$.

Lemma 4. *The optimum utility of $\mathcal{S}_k$ is greater than $D_K - 1$.*

Proof. Construct a solution x as follows. Let $D = \max\{D_I, D_K + 1\}$. If the distance between the agent v and the objective k in $\mathcal{G}_k$ is $4j + 1$ for some j, set $x_v = 1 - 1/D^{2j+1}$. If the distance is $4j + 3$, set $x_v = 1/D^{2j+2}$.

To see that x is a feasible solution, first observe that feasibility is clear for a leaf constraint. Any non-leaf constraint $i \in I$ has at most D_I neighbours, and the distance between k and i is $4j + 2$ for some j. Thus

$$\sum_{v \in V_i} a_{iv} x_v \le 1 - 1/D^{2j+1} + (D_I - 1)/D^{2j+2} < 1.$$

Let ω_k be the utility of x. We show that $\omega_k > D_K - 1$. First, consider the objective k. We have

$$\sum_{v \in V_k} c_{kv} x_v = D_K(1 - 1/D) > D_K - 1.$$

Second, each objective $h \in K' \setminus \{k\}$ has D_K neighbours and the distance between h and k is $4j$ for some j. Thus

$$\sum_{v \in V_h} c_{hv} x_v = 1/D^{2j} + (D_K - 1)(1 - 1/D^{2j+1}) > D_K - 1.$$

Finally, each objective $h \in K \setminus K'$ has 2 neighbours and the distance between h and k is $4j$ for some j; the coefficients are $c_{hv} = D_K - 1$. Thus

$$\sum_{v \in V_h} c_{hv} x_v = (D_K - 1)(1/D^{2j} + 1 - 1/D^{2j+1}) > D_K - 1. \qquad \square$$

4.3   Proof of Theorem 2

Let $\Delta_I \ge 2$ and $\Delta_K \ge 2$. Assume that $\mathcal{A}$ is a local approximation algorithm with the approximation ratio α. Set $D_I = \Delta_I$, $D_K = \Delta_K$ and $s = 0$. Let r be the local horizon of the algorithm, rounded up to a multiple of 4. Construct the instance $\mathcal{S}$ as described in Sect. 4.1; it is a 0/1 bipartite max-min LP, and it satisfies the degree bounds Δ_I and Δ_K. Apply the algorithm $\mathcal{A}$ to $\mathcal{S}$. The algorithm produces a feasible solution x. By Lemma 3 there is a constraint k such that $\sum_{v \in V_k} x_v \le \Delta_K/\Delta_I$.

Now construct $\mathcal{S}_k$ as described in Sect. 4.2; this is another 0/1 bipartite max-min LP. Apply $\mathcal{A}$ to $\mathcal{S}_k$. The algorithm produces a feasible solution x'. The radius r neighbourhoods of the agents $v \in V_k$ are identical in $\mathcal{S}$ and $\mathcal{S}_k$; therefore the algorithm must make the same decisions for them, and we have $\sum_{v \in V_k} x'_v \le \Delta_K/\Delta_I$. But by Lemma 4 there is a feasible solution of $\mathcal{S}_k$ with utility greater than $\Delta_K - 1$; therefore the approximation ratio of $\mathcal{A}$ is $\alpha > (\Delta_K - 1)/(\Delta_K/\Delta_I)$. This completes the proof of Theorem 2.

4.4 Proof of Theorem 3

Let $\Delta_I \geq 3$, $\Delta_K \geq 3$, and $0 < \delta < 1/10$. Assume that $\mathcal{A}$ is a local approximation algorithm with the approximation ratio α. Set $D_I = 3$, $D_K = 3$, and $s = \lceil 4/(7\delta) - 1/2 \rceil$. Let r be the local horizon of the algorithm, rounded up to a multiple of 4.

Again, construct the instance $\mathcal{S}$. The relative growth of $\mathcal{G}$ is at most $1 + 2^j/((2^j - 1)(2s + 1))$ beyond radius $R = j(4s + 2)$; indeed, each set of 2^j new agents can be accounted for $1 + 2 + \cdots + 2^{j-1} = 2^j - 1$ chains with $2s + 1$ agents each. Choosing $j = 3$, the relative growth of $\mathcal{G}$ is at most $1 + \delta$ beyond radius R.

Apply $\mathcal{A}$ to $\mathcal{S}$. By Lemma 3 we know that there exists an objective h such that $\sum_{v \in V_h} x_v \leq 2 - 2/(3s + 2)$. Choose a $k \in K'$ nearest to h. Construct $\mathcal{S}_k$ and apply $\mathcal{A}$ to $\mathcal{S}_k$. The local neighbourhoods of the agents $v \in V_h$ are identical in $\mathcal{S}$ and $\mathcal{S}_k$. By Lemma 4 there is a feasible solution of $\mathcal{S}_k$ with utility greater than 2. Using the assumption $\delta < 1/10$, we obtain

$$\alpha > \frac{2}{2 - 2/(3s + 2)} = 1 + \frac{1}{3s + 1} \geq 1 + \frac{1}{3(4/(7\delta) + 1/2) + 1} > 1 + \frac{\delta}{2}.$$

This completes the proof of Theorem 3.

Acknowledgements

We thank anonymous reviewers for their helpful feedback. This research was supported in part by the Academy of Finland, Grants 116547 and 117499, and by Helsinki Graduate School in Computer Science and Engineering (Hecse).

References

1. Naor, M., Stockmeyer, L.: What can be computed locally? SIAM Journal on Computing 24(6), 1259–1277 (1995)
2. Floréen, P., Hassinen, M., Kaski, P., Suomela, J.: Local approximation algorithms for a class of 0/1 max-min linear programs (manuscript, 2008) arXiv:0806.0282 [cs.DC]
3. Floréen, P., Kaski, P., Musto, T., Suomela, J.: Approximating max-min linear programs with local algorithms. In: Proc. 22nd IEEE International Parallel and Distributed Processing Symposium (IPDPS), Miami, FL, USA. IEEE, Piscataway (2008)
4. Linial, N.: Locality in distributed graph algorithms. SIAM Journal on Computing 21(1), 193–201 (1992)
5. Angluin, D.: Local and global properties in networks of processors. In: Proc. 12th Annual ACM Symposium on Theory of Computing (STOC), Los Angeles, CA, USA, pp. 82–93. ACM Press, New York (1980)
6. Papadimitriou, C.H., Yannakakis, M.: Linear programming without the matrix. In: Proc. 25th Annual ACM Symposium on Theory of Computing (STOC), San Diego, CA, USA, pp. 121–129. ACM Press, New York (1993)

7. Kuhn, F., Moscibroda, T., Wattenhofer, R.: The price of being near-sighted. In: Proc. 17th Annual ACM-SIAM Symposium on Discrete Algorithms (SODA), Miami, FL, USA, pp. 980–989. ACM Press, New York (2006)
8. Lynch, N.A.: A hundred impossibility proofs for distributed computing. In: Proc. 8th Annual ACM Symposium on Principles of Distributed Computing (PODC), Edmonton, Canada, pp. 1–28. ACM Press, New York (1989)
9. Kuhn, F., Moscibroda, T.: Distributed approximation of capacitated dominating sets. In: Proc. 19th Annual ACM Symposium on Parallel Algorithms and Architectures (SPAA), San Diego, CA, USA, pp. 161–170. ACM Press, New York (2007)
10. Kuhn, F., Moscibroda, T., Wattenhofer, R.: On the locality of bounded growth. In: Proc. 24th Annual ACM Symposium on Principles of Distributed Computing (PODC), Las Vegas, NV, USA, pp. 60–68. ACM Press, New York (2005)
11. Lubotzky, A., Phillips, R., Sarnak, P.: Ramanujan graphs. Combinatorica 8(3), 261–277 (1988)
12. Lazebnik, F., Ustimenko, V.A.: Explicit construction of graphs with an arbitrary large girth and of large size. Discrete Applied Mathematics 60(1–3), 275–284 (1995)
13. Hoory, S.: On Graphs of High Girth. PhD thesis, Hebrew University, Jerusalem (March 2002)
14. McKay, B.D., Wormald, N.C., Wysocka, B.: Short cycles in random regular graphs. Electronic Journal of Combinatorics 11(1), R66 (2004)
15. Kuhn, F., Moscibroda, T., Wattenhofer, R.: What cannot be computed locally! In: Proc. 23rd Annual ACM Symposium on Principles of Distributed Computing (PODC), St. John's, Newfoundland, Canada, pp. 300–309. ACM Press, New York (2004)
16. Anderson, E.J., Nash, P.: Linear Programming in Infinite-Dimensional Spaces: Theory and Applications. John Wiley & Sons, Ltd, Chichester (1987)
17. Godsil, C., Royle, G.: Algebraic Graph Theory. Graduate Texts in Mathematics, vol. 207. Springer, New York (2004)
18. Hocking, J.G., Young, G.S.: Topology. Addison-Wesley, Reading (1961)
19. Munkres, J.R.: Topology, 2nd edn. Prentice-Hall, Upper Saddle River (2000)
20. Amit, A., Linial, N., Matoušek, J., Rozenman, E.: Random lifts of graphs. In: Proc 12th Annual ACM-SIAM Symposium on Discrete Algorithms (SODA), Washington, DC, USA, pp. 883–894. Society for Industrial and Applied Mathematics, Philadelphia (2001)
21. Esparza, J., Heljanko, K.: A new unfolding approach to LTL model checking. In: Welzl, E., Montanari, U., Rolim, J.D.P. (eds.) ICALP 2000. LNCS, vol. 1853, pp. 475–486. Springer, Heidelberg (2000)

Minimizing Average Flow Time
in Sensor Data Gathering*

Vincenzo Bonifaci[1,3], Peter Korteweg[2],
Alberto Marchetti-Spaccamela[3,**], and Leen Stougie[2,4,***]

[1] Università degli Studi dell'Aquila, Italy
bonifaci@dis.uniroma1.it
[2] Eindhoven University of Technology, The Netherlands
p.korteweg@tue.nl, l.stougie@tue.nl
[3] Sapienza Università di Roma, Italy
alberto@dis.uniroma1.it
[4] CWI, Amsterdam, The Netherlands
stougie@cwi.nl

Abstract. Building on previous work [Bonifaci et al., *Minimizing flow time in the wireless gathering problem*, STACS 2008] we study data gathering in a wireless network through multi-hop communication with the objective to minimize the average flow time of a data packet. We show that for any $\epsilon \in (0, 1)$ the problem is NP-hard to approximate within a factor better than $\Omega(m^{1-\epsilon})$, where m is the number of data packets. On the other hand, we give an online polynomial time algorithm that we analyze using resource augmentation. We show that the algorithm has average flow time bounded by that of an optimal solution when the clock speed of the algorithm is increased by a factor of five. As a byproduct of the analysis we obtain a 5-approximation algorithm for the problem of minimizing the average completion time of data packets.

1 Introduction

In this paper we study a scheduling problem motivated by data gathering in sensor networks: we are given a graph where nodes represent sensors (or wireless stations), edges possible communication links and there is special node, the base station (also called the *sink*). Over time events occur at the nodes; events are unpredictable and each such event triggers the invoice of a packet toward the sink using edges of the graph (multihop communication). The goal is to find an on-line schedule (i.e. a schedule unaware of future events) that optimizes a given objective function; the obtained schedule must comply with constraints posed by

---

* Research supported by EU FET-project under contract no. FP6-021235-2 ARRIVAL and by the EU COST-action 293 GRAAL.
** Research supported by EU ICT-FET 215270 FRONTS and MIUR-FIRB Italy-Israel project RBIN047MH9.
*** Research supported the Dutch BSIK-BRICKS project.

S. Fekete (Ed.): ALGOSENSORS 2008, LNCS 5389, pp. 18–29, 2008.

interferences in the communication that restrict contemporary communication between nearby nodes.

The problem was introduced in [4] in the context of wireless access to the Internet in villages. For a motivation and history of the problem we refer to [4, 6] or to the PhD-thesis of Korteweg [14]. Here we restrict to explaining the ingredients.

In our model we assume that sensor nodes share a common clock, thus allowing division of time into rounds. At each round a node can either send a packet or receive a packet or be silent. Since not all nodes in the network can communicate with each other directly, packets have to be sent through several intermediate nodes before they can be gathered at the sink through multihop communication.

The key issue is interference. The model we use was proposed by Bermond et al. in [4]: there is an edge between nodes i and j if corresponding sensors can directly communicate (i.e. they are within transmission range of each other). An integer parameter d_I models the interference radius, with distance between any pair of vertices expressed as the number of edges on a shortest path between them. A node j successfully receives a packet from one of its neighbors if no other node within distance d_I from j is transmitting in the same round. In fact, Bermond et al. in [4] proposed an integer transmission radius $d_T \leq d_I$, indicating the maximum number of edges between any two consecutive hops for every message. In that sense we just consider the case $d_T = 1$.

Given an instance of the data gathering problem several possible objective functions can be considered. In [4] the authors considered the goal of minimizing the completion time (makespan) of the schedule. Makespan is prominently used for assessing the performance of scheduling algorithms; however it is now accepted that it is an unsuitable measure when jobs arrive in continuous streams [3].

Today, flow time minimization is a largely used criterion in scheduling theory that more suitably allows to assess the quality of service provided when multiple requests occur over time [8, 9, 13, 19]. The flow time of a data packet is the time elapsed between its release at the release node and its arrival at the sink. In [6] the considered objective was to minimize the maximum flow time of a data packet. Here we study the problem of minimizing the average flow time or total flow time of data packets.

Both flow time objective functions have been thoroughly studied by the scheduling community and there is an extensive literature both for the on-line and off-line algorithms for which we refer to [18]. Here we remark that the two problems have fundamentally different characteristics and that results for one problem do not carry over to the other one. In general minimizing total flow time appears to be a more difficult problem than minimizing max flow time.

In fact, if we consider on-line algorithms and if the objective function requires to minimize the maximum flow time then the *First In Firts Out (FIFO)* heuristic is the natural choice: at each time FIFO schedules the earliest released jobs among unfinished jobs. In the case of uniprocessor scheduling FIFO produces an optimal solution while in the case of parallel machines it gives a $3 - 2/m$

approximation (where m denotes the number of used machines) [3]. When the objective function is total flow time the natural heuristic to be used is *Shortest Remaining Processing Time (SRPT)* first, the on-line strategy that schedules first jobs closer to completion. This heuristic is optimal in the case of one machine but it is not constant approximate in the case of parallel machines. In fact SRPT gives a solution that is $\Theta(\min(\log \frac{n}{m}, \log P))$ approximate, where n and m denote respectively the number of jobs and the number of machines and P denotes the ratio between the longest and the smallest processing time [17]. In the same paper it is shown that no on-line randomized algorithm can achieve a better competitive ratio.

We resume the problem that we study in the following definition. A mathematical formalization will be given in Section 2.

F-WGP. An instance of the *Wireless Gathering Problem* (WGP) is given by a network which consists of several stations (nodes) and one base station (the sink), modeled as a graph, together with the interference radius d_I; over time data packets arrive at stations that have to be gathered at the base station.

A feasible solution of an instance of WGP is a schedule without interference which determines for each packet both route and times at which it is sent.

The objective is to minimize the average flow time of packets.

Having defined the problem we now discuss two key aspects that restrict the class of algorithms that we consider. Firstly, we are interested in on-line algorithms. At time t an on-line algorithm makes its decisions on events that occur at or before t and it ignores future events. Competitive analysis compares the solution of the on-line algorithm with the optimal solution obtained by an omniscient adversary. We refer the reader to [7] for a comprehensive survey on on-line algorithms and competitive analysis. Secondly we restrict to simple distributed algorithms that might be amenable for implementation or that faithfully represent algorithms used in practice. In fact, we think that sophisticated algorithms are impractical for implementations and have mainly theoretical interest.

Related Work. The Wireless Gathering Problem was introduced by Bermond et al. [4] in the context of wireless access to the Internet in villages. The authors proved that the problem of minimizing the completion time is NP-hard and presented a greedy algorithm with asymptotic approximation ratio at most 4. They do not consider release times. In [5] we considered the same problem with arbitrary release times and proposed a simple on-line greedy algorithm with the same approximation ratio. Both papers do not consider distributed algorithms. The present paper builds on [6] in which on-line distributed algorithms are analysed for the problem when the objective is to minimize the maximum flow time of a data packet.

The case $d_I = 1$ has been extensively considered (see for example [2, 11, 12]); we remark that assuming $d_I = 1$ or assuming that interferences/transmissions are modeled according to the well known unit disk graph model *does not* adequately represent interferences as they occur in practice [21]. We also observe that almost all of the previous literature considered the objective of minimizing the completion time (see for example [1, 2, 4, 11, 12, 15, 20]).

Finally, we note that many papers study broadcasting in wireless networks [1, 20]. However, we stress that broadcasting requires to broadcast the same piece of information to all nodes; so the two problems are intrinsically different. In particular, given a broadcast schedule it is not possible to obtain a gathering schedule by simply exchanging sender and receiver. This would only be true if data packets could be aggregated into single packets and disaggregated afterwards, or in case of private broadcasting in which each data packet has a specific recipient address.

Results of the Paper. The F-WGP problem is NP-hard, as can be shown by using a modification of a construction by Bermond et al. [4] (this also implies that C-WGP is NP-hard). In Section 3.2 we show the stronger result that F-WGP is also hard to approximate, namely for any $\epsilon \in (0, 1)$ there is a lower bound of $\Omega(m^{1-\epsilon})$ on the approximation ratio (m is the number of packets). We notice that this is a stronger inapproximability result than the one we obtained for the maximum flow time minimization problem [6]. The construction of both results have many similarities though, both being based on the same reduction. We will point out the differences between the two in the analysis in Section 3.2.

In Section 3.1 we propose an online polynomial time algorithm based on the Shortest Remaining Processing Time first rule. We show that it yields a pseudoapproximation to F-WGP, in the sense that its average flow time is not larger than that of an optimal solution, assuming that the algorithm runs at speed 5 times higher than an optimal algorithm. This type of analysis, called *resource augmentation*, has already been used successfully in the context of many machine scheduling problems [9, 13]. We showed already in [6] that resource augmentation is also a useful tool for the analysis of algorithms for wireless communication, allowing to obtain positive results for data gathering problems.

It is not surprising that a FIFO-type algorithm as studied for minimizing maximum flow time in [6] does not work for minimizing average flow time when SRPT rule is the one to use. However we observe that we are unable to prove our result for SRPT but only for a modified rule. It remains an interesting open problem to decide whether a similar result can be proved for SRPT.

As a byproduct of our analysis we also obtain an online, polynomial time 5-approximation algorithm for C-WGP. An additional useful property of our algorithms is that nodes only need a limited amount of information in order to coordinate.

2   Mathematical Formulations

In this section we define the problem more formally. The model we use is not new: it can be seen as a generalization of a well-studied model for packet radio networks [1, 2]. It has also been used in more recent work [4, 6]. We summarize it for independent reading.

An instance of WGP consists of a graph $G = (V, E)$, a *sink* node $s \in V$, a positive integer d_I, and a set of *data packets* $J = \{1, 2, \ldots, m\}$. Each packet $j \in J$ has a release node or *origin* $o_j \in V$ and a *release date* $r_j \in \mathbb{R}_+$. The release date specifies the time at which the packet enters the network, i.e. packet j is not *available* for transmission before round r_j.

Time is slotted; each time slot is called a *round*. The rounds are numbered $0, 1, 2, \ldots$ During each round a node may either be *sending* a packet, be *receiving* a packet or be inactive. If two nodes u and v are adjacent, then u can send a packet to v during a round. If node u sends a packet j to v in some round, the pair (u, v) is said to be a *call* from u to v. For each pair of nodes $u, v \in V$, the *distance* between u and v, denoted by $d(u, v)$, is the minimum number of edges between u and v in G. Two calls (u, v) and (u', v') *interfere* if they occur in the same round and either $d(u', v) \leq d_I$ or $d(u, v') \leq d_I$; otherwise the calls are *compatible*. The parameter d_I is called the *interference radius*.

We formulate our problem as an offline problem, but the algorithms we analyze are online, in the sense that when scheduling a certain round they only use the information about packets released not later than the same round.

A solution for a WGP instance is a schedule of compatible calls such that all packets are ultimately collected at the sink. Since it suffices to keep only one copy of each packet during the execution of a schedule, we assume that at any time there is a unique copy of each packet. Packets cannot be aggregated in this model.

Given a schedule, let v_j^t be the unique node holding packet j at time t. The value $C_j := \min\{t : v_j^t = s\}$ is called the *completion time* of packet j, while $F_j := C_j - r_j$ is the *flow time* of packet j. In this paper we are interested in the minimization of $\sum_j F_j$ (F-WGP). As a byproduct of the analysis of F-WGP, we also give a result on the minimization of $\sum_j C_j$ (C-WGP).

Some additional notation: we denote by $\delta_j := d(o_j, s)$ the minimum number of calls required for packet j to reach s. We also define $\gamma := d_I + 2$, which is a lower bound, because of interference, on the inter arrival time at s of two messages that use the same $d_I + 2$ nodes as hops on their way to s. The *critical region* is the set $\{v \in V \mid d(s, v) \leq \lfloor (d_I - 1)/2 \rfloor\}$, which is the region around s in which no two nodes can receive a message in the same round. Related to this region we define $\gamma^* := \lfloor (d_I + 1)/2 \rfloor$, which is then a lower bound, because of interference, on the inter arrival time at s between any two messages that are released outside the critical region.

In what follows we assume that the reader is familiar with the basic notions related to approximation algorithms. We also use resource augmentation to assess our algorithms. We consider augmentation based on speed, meaning that the algorithm can schedule compatible calls with higher speed than an optimal algorithm. For any $\sigma \geq 1$, we call an algorithm a σ-speed algorithm if the time used by the algorithm to schedule a set of compatible calls is $1/\sigma$ time units. Thus, the ith round occurs during time interval $[i/\sigma, (i+1)/\sigma)$. We notice that the release dates of packets are independent of the value of σ.

3 Gathering to Minimize Average Flow Time

3.1 The Interleaved Shortest Remaining Processing Time Algorithm

We introduce an algorithm that we call INTERLEAVED SRPT and prove that a constant-factor speed augmentation is enough to enable this algorithm to outperform the optimal average flow time of the original instance. The algorithm is based on a well-known scheduling algorithm, the shortest remaining processing time first rule (SRPT) [22], so we first describe this algorithm in the context of WGP.

Algorithm 1. SHORTEST REMAINING PROCESSING TIME (SRPT)

---

for $k = 0, 1, 2, \ldots$ **do**
  At time $t = k/\sigma$, let $1, \ldots, m'$ be the available packets in order of non-decreasing distance to the sink (that is, $d(v_1^t, s) \le d(v_2^t, s) \le \ldots \le d(v_{m'}^t, s)$)
  for $j = 1$ to m' **do**
    Send j to the next hop along an arbitrary shortest path from v_j^t to the sink, unless this creates interference with a packet j' with $j' < j$
  end for
end for

Every iteration k in the algorithm corresponds to a round of the schedule. We notice that this algorithm is a dynamic-priority algorithm, in the sense that the ordering in which packets are scheduled can change from round to round. We also notice that, $\delta_j < \gamma^*$ for each packet $j \in J$ (that is, when all packets are released inside the critical region), then WGP reduces to a single machine scheduling problem with preemption. The problem of minimizing average flow-times is then equivalent to the single machine scheduling problem with the same objective, allowing preemption and jobs having release times: $1|r_j, \text{pmtn}| \sum_j F_j$ in terms of [16]. For the off-line problem minimizing average flow-time has the same optimal solution as minimizing average completion times. Schrage [22] showed that SRPT solves the latter problem to optimality, which motivated our use of SRPT.

Consider a schedule generated by σ-speed SRPT, that is, every round is executed in time $1/\sigma$. It will be convenient to refer to round $[i/\sigma, (i+1)/\sigma)$ as "round i/σ". Recall that we use C_j to denote the completion time of packet j.

We denote the ith packet to arrive at the sink in this schedule as $p(i)$, for $1 \le i \le m$. We define a *component* as a set S of packets with the following properties:

1. There is an index a such that $S = \{p(a), p(a+1), \ldots, p(a + |S| - 1)\}$;
2. If $i \ge 1$ and $i \le |S| - 1$, then $C_{p(a+i)} \le C_{p(a+i-1)} + \gamma/\sigma$;
3. If $a + |S| \le m$, then $C_{p(a+|S|)} > C_{p(a+|S|-1)} + \gamma/\sigma$.

That is, a component is a maximal set of packets arriving subsequently at the sink, each within time γ/σ of the previous packet. It follows from the definition that the set J of all packets can be partitioned into components $T_1, \ldots, T_\ell$, for some ℓ.

Lemma 1. *For any component T we have* $\min_{j \in T} C_j = \min_{j \in T}(r_j + \delta_j/\sigma)$.

Proof. Consider the partition of the packet set J into components $T_1, \ldots, T_\ell$. The components are ordered so that $\max_{j \in T_i} C_j < \min_{k \in T_{i+1}} C_k$ for each i; by definition of a component such an ordering exists.

Let $S(i) = \cup_{h=i}^\ell T_h$, for $1 \le i \le \ell$. We define $\bar{t}_i := \min_{j \in S(i)}(r_j + \delta_j/\sigma)$, the earliest possible arrival time of any packet in $S(i)$, and $\underline{t}_i := \max\{r_j : j \in S(i) \text{ and } r_j + \delta_j/\sigma = \bar{t}_i\}$, the maximum release date of a packet in $S(i)$ with earliest possible arrival time $\bar{t}_i$. Consider the following set of packets, for $\underline{t}_i \le t \le \bar{t}_i$:

$$M_i(t) = \{j \in S(i) : r_j \le t \text{ and } d(v_j^t, s) \le d(v_k^t, s) \text{ for all } k \in S(i)\},$$

Note that $|M_i(t)| \ge 1$ for $\underline{t}_i \le t \le \bar{t}_i$, because no packet in $S(i)$ arrives at the sink before round $\bar{t}_i$. The crucial observation is that for each round t we have that if no packet in $M_i(t)$ is sent towards the sink, then some packet in $J \setminus S(i)$ is sent; also, by definition of SRPT this packet must be closer to the sink during round t than any packet in $M_i(t)$. The proof of the lemma follows from the following claim.

Claim. *For $i = 1, 2, \ldots, \ell$ and for any round $t \in [\underline{t}_i, \bar{t}_i]$ there exists $j \in M_i(t)$ such that $t + d(v_j^t, s)/\sigma \le \bar{t}_i$.*

Suppose the claim holds. Then choosing $t = \bar{t}_i$ implies that for each $i, 1 \le i \le \ell$, there is a packet $j \in M_i(\bar{t}_i)$ which arrives at the sink in round $\bar{t}_i = \min_{k \in S(i)} r_k + \delta_k/\sigma$. As a consequence $j \in T_i$, and $\bar{t}_i = r_j + \delta_j/\sigma$, which proves the lemma. □

Proof of Claim. The claim trivially holds for $t = \underline{t}_i$, because some packet $j \in S(i)$ with earliest possible arrival time $\bar{t}_i$ is released in round $\underline{t}_i$, hence $\underline{t}_i + d(v_j^{\underline{t}_i}, s)/\sigma = r_j + \delta_j/\sigma = \bar{t}_i$.

First, consider the case $i = 1$; then $S(1) = J$. It follows from the observation above and the facts $J \setminus S(1) = \emptyset$ and $|M_1(t)| \ge 1$ for $t \in [\underline{t}_1, \bar{t}_1]$ that during each round $t \in [\underline{t}_1, \bar{t}_1]$ some packet in $M_1(t)$ is sent towards the sink. This proves the claim for this case.

Next consider $i > 1$. Suppose that during each round $t \in [\underline{t}_i, \bar{t}_i]$ some packet in $M_i(t)$ is sent towards the sink. Then as above this would prove the claim. Otherwise, there must be a maximal round $t' \in [\underline{t}_i, \bar{t}_i]$ in which no packet in $M_i(t')$ is sent towards the sink. By definition of SRPT there is a packet $k \in J \setminus S(i)$ which is sent, and a packet $j \in M_i(t')$ for which $d(v_j^{t'}, v_k^{t'}) \le d_I + 1$. Since j is not sent during round t', we also have $d(v_j^{t'+1/\sigma}, v_k^{t'}) \le d_I + 1$. Additionally, $d(v_k^{t'}, s)/\sigma \le C_k - t'$ because otherwise k could not reach the sink by time C_k. Now for each round $t \in [t' + 1/\sigma, \bar{t}_i]$ a packet in $M_i(t)$ is sent. In particular, there must be a packet from the set $\cup_{t \in (t', \bar{t}_i]} M_i(t)$, call it q, that arrives at the sink no later than j would arrive if j were always sent from round $t' + 1/\sigma$ on. We have

$$C_q \leq (t' + 1/\sigma) + \left(d(v_j^{t'+/\sigma}, v_k^{t'}) + d(v_k^{t'}, s)\right)/\sigma \leq C_k + (d_I + 2)/\sigma = C_k + \gamma/\sigma.$$

That is, packet $q \in S(i)$ arrives at most γ/σ time units after packet $k \in J \setminus S(i)$, which contradicts the fact that q and k are in different components. Thus this case never occurs and the claim holds.   □

We now describe INTERLEAVED SRPT. The algorithm partitions the set of packets J in two subsets, $J^{\text{in}} := \{j \in J : \delta_j < \gamma^*\}$ and $J^{\text{out}} := \{j \in J : \delta_j \geq \gamma^*\}$. The two subsets are scheduled in an interleaved fashion using SRPT. The pseudocode is given as Algorithm 2.

Algorithm 2. INTERLEAVED SRPT (ISRPT)

Inizialization: $c := 1$
loop
   if $c \neq 0$ (mod 5) **then**
      execute one round of SRPT on the set J^{out}
   else
      execute one round of SRPT on the set J^{in}
   end if
   $c := c + 1$
end loop

In the performance analysis of INTERLEAVED SRPT we use the following lower bound on the sum of optimal completion times of a subset of jobs in J^{out}, which is obtained as a direct corollary of Lemma 2 in [5].

Lemma 2. *Let $S \subseteq J^{\text{out}}$. If C_j^* denotes the completion time of packet j in any feasible schedule, we have*

$$\sum_{j \in S} C_j^* \geq \sum_{i=0}^{|S|-1} (C_S + i\gamma^*)$$

where $C_S := \min_{j \in S} r_j + \delta_j$.   □

Theorem 1. 5-*speed ISRPT is optimal for* F-WGP.

Proof. Let C_j be the completion time of packet j in a 5-speed ISRPT schedule, and let C_j^* be the completion time of packet j in any feasible (possibly optimal) 1-speed schedule. We prove the theorem by showing that $\sum_{j \in J^{\text{in}}} C_j \leq \sum_{j \in J^{\text{in}}} C_j^*$ and $\sum_{j \in J^{\text{out}}} C_j \leq \sum_{j \in J^{\text{out}}} C_j^*$.

Consider first the packets in J^{in}. Since we are executing ISRPT at speed 5, and the set J^{in} is considered once every five iterations, we have that every one time unit a round of SRPT is executed on the set J^{in}. So the completion times of the packets in J^{in} are not worse than those that would be obtained by running SRPT with unit speed on J^{in} alone. On the other hand, inside the critical region the gathering problem is nothing else than the scheduling problem $1|r_j, \text{pmtn}| \sum_j C_j$, meaning that SRPT is optimal. It follows that $\sum_{j \in J^{\text{in}}} C_j \leq \sum_{j \in J^{\text{in}}} C_j^*$.

Consider now the packets in J^{out}. Because the *first* four out of every five rounds of ISRPT this set is scheduled using SRPT, the completion time of each packet in J^{out} is not larger than the completion time of the same packet in a 4-speed SRPT schedule of J^{out}:

$$C_j \leq \overline{C}_j \text{ for all } j \in J^{\text{out}}, \tag{1}$$

where $\overline{C}_j$ is the completion time of j in a 4-speed SRPT schedule of J^{out}. Consider any component T in this latter schedule. By Lemma 1,

$$\sum_{j \in T} \overline{C}_j \leq \sum_{0 \leq i < |T|} (C_T + \frac{1}{4} i \gamma). \tag{2}$$

On the other hand by Lemma 2,

$$\sum_{j \in T} C_j^* \geq \sum_{0 \leq i < |T|} (C_T + i \gamma^*). \tag{3}$$

So, since $\gamma \leq 4\gamma^*$ for every value of d_I, it follows by combining (1), (2) and (3) that $\sum_{j \in T} C_j \leq \sum_{j \in T} C_j^*$. The result follows by summing over all the components. □

Corollary 1. *There is an online 5-approximation algorithm for* C-WGP.

Proof. Notice that the analysis of the above theorem also yields that

$$\sum_{j \in J} C_j \leq \sum_{j \in J} C_j^*,$$

where C_j is the completion time of packet j in the schedule generated by 5-ISRPT. We can now simulate the schedule generated by 5-ISRPT by running it at a lower speed: whatever 5-ISRPT does at time t, a unit-speed algorithm can do at time $5t$. The schedule can be constructed online and clearly it respects the release dates. If C_j' is the completion time of packet j in the new schedule, we obtain

$$\sum_{j \in M} C_j' = \sum_{j \in M} 5 \cdot C_j \leq 5 \sum_{j \in M} C_j^*.$$

□

3.2   Approximation Hardness

In this section we show that, for any $\epsilon \in (0,1)$, no polynomial time algorithm can approximate F-WGP within a factor $\Theta(m^{1-\epsilon})$ unless P=NP. This explains why resource augmentation is necessary to obtain a useful bound.

The proof builds upon the one in our previous paper [6] and is based on the hardness of the *induced matching* problem. A matching M in a graph G is an *induced matching* if no two edges in M are joined by an edge of G.

INDUCED BIPARTITE MATCHING (IBM)
Instance: a bipartite graph G and an integer k.
Question: does G have an induced matching of size at least k?

The optimization version of the above problem is hard to approximate: there exists an $\alpha > 1$ such that it is NP-hard to distinguish between graphs with induced matchings of size k and graphs in which all induced matchings are of size at most k/α [10].

Theorem 2. *Let* $\epsilon \in (0,1)$. *Unless* P=NP, *no polynomial time algorithm can approximate* F-WGP *within a ratio better than* $\Omega(m^{1-\epsilon})$.

Proof. We only describe the additional steps needed with respect to the proof of inapproximability for minimizing maximum flow time, Theorem 3.2 in [6]. As shown in [6], it is possible to construct in polynomial time, given an IBM instance I, an instance I' of WGP with an arbitrary number m of packets such that the following hold:

1. if I has an induced matching of size k, then there is a schedule for I' with *maximum* flow time $2k + 1$;
2. if all induced matchings of I are of size at most k/α, then in every schedule for I' there will be a round in which $\Theta(m/k)$ packets have been released but not yet collected at the sink.

In the above construction choose $m := (1 - 1/\alpha)^{-1}(1 + k/\alpha)(2k + 1)k^{3/\epsilon - 2} = \Theta(k^{3/\epsilon})$. In case (1), since there is a schedule for I' in which the maximum flow time is $2k + 1$, we have that in the same schedule the *total* flow time is bounded by $(2k + 1) \cdot m = O(m^{1+\epsilon/3})$.

Instead, in case (2), we have that in any schedule there will be a round when $\Theta(m/k)$ packets are available but not yet delivered. We now use the simple fact that if at any time during a schedule there are p available packets that still need to reach the sink, then the total flow time of the schedule is $\Omega(p^2)$; this is true because the sink can only absorb at most one packet per round. Since $p = \Theta(m/k)$, it follows that the total flow time is $\Omega(m^2/k^2) = \Omega(m^{2-2\epsilon/3})$.

The ratio between the total flow time achievable in cases (2) and (1) is $\Omega(m^{1-\epsilon})$. Thus, any polynomial-time algorithm approximating the total flow time within a better ratio could be used to approximate IBM within factor α, which is NP-hard. $\square$

References

[1] Bar-Yehuda, R., Goldreich, O., Itai, A.: On the time-complexity of broadcast in multi-hop radio networks: an exponential gap between determinism and randomization. Journal of Computer and Systems Sciences 45(1), 104–126 (1992)

[2] Bar-Yehuda, R., Israeli, A., Itai, A.: Multiple communication in multihop radio networks. SIAM Journal on Computing 22(4), 875–887 (1993)

[3] Bender, M.A., Chakrabarti, S., Muthukrishnan, S.: Flow and stretch metrics for scheduling continuous job streams. In: Proc. 9th Symp. on Discrete Algorithms, pp. 270–279. SIAM, Philadelphia (1998)

[4] Bermond, J., Galtier, J., Klasing, R., Morales, N., Pérennes, S.: Hardness and approximation of gathering in static radio networks. Parallel Processing Letters 16(2), 165–183 (2006)

[5] Bonifaci, V., Korteweg, P., Marchetti-Spaccamela, A., Stougie, L.: An approximation algorithm for the wireless gathering problem. In: Arge, L., Freivalds, R. (eds.) SWAT 2006. LNCS, vol. 4059, pp. 328–338. Springer, Heidelberg (2006)

[6] Bonifaci, V., Korteweg, P., Marchetti-Spaccamela, A., Stougie, L.: Minimizing flow time in the wireless gathering problem. In: Proc. 25th Symp. on Theoretical Aspects of Computer Science, pp. 109–120. IBFI Dagstuhl (2008)

[7] Borodin, A., El-Yaniv, R.: Online Computation and Competitive Analysis. Cambridge University Press, Cambridge (1998)

[8] Chan, H.-L., Lam, T.W., Liu, K.-S.: Extra unit-speed machines are almost as powerful as speedy machines for competitive flow time scheduling. In: Proc. 17th Symp. on Discrete Algorithms, pp. 334–343. SIAM, Philadelphia (2006)

[9] Chekuri, C., Goel, A., Khanna, S., Kumar, A.: Multi-processor scheduling to minimize flow time with epsilon resource augmentation. In: Proc. 36th Symp. on Theory of Computing, pp. 363–372. ACM, New York (2004)

[10] Duckworth, W., Manlove, D., Zito, M.: On the approximability of the maximum induced matching problem. Journal of Discrete Algorithms 3(1), 79–91 (2005)

[11] Florens, C., Franceschetti, M., McEliece, R.J.: Lower bounds on data collection time in sensory networks. IEEE Journal on Selected Areas in Communications 22, 1110–1120 (2004)

[12] Gargano, L., Rescigno, A.A.: Optimally fast data gathering in sensor networks. In: Královič, R., Urzyczyn, P. (eds.) MFCS 2006. LNCS, vol. 4162, pp. 399–411. Springer, Heidelberg (2006)

[13] Kalyanasundaram, B., Pruhs, K.: Speed is as powerful as clairvoyance. Journal of the ACM 47(4), 617–643 (2000)

[14] Korteweg, P.: Online gathering algorithms for wireless networks. PhD thesis, Technische Universiteit Eindhoven (2008)

[15] Anil Kumar, V.S., Marathe, M.V., Parthasarathy, S., Srinivasan, A.: End-to-end packet-scheduling in wireless ad-hoc networks. In: Munro, J.I. (ed.) Proc. 15th Symp. on Discrete Algorithms, pp. 1021–1030. SIAM, Philadelphia (2004)

[16] Lageweg, B.J., Lawler, E.L., Lenstra, J.K., Rinnooy Kan, A.H.G.: Computer-aided complexity classification of combinatorial problems. Communications of the ACM 25, 817–822 (1982)

[17] Leonardi, S., Raz, D.: Approximating total flow time on parallel machines. Journal of Computer and Systems Sciences 73(6), 875–891 (2007)

[18] Leung, J.Y.-T. (ed.): Handbook of Scheduling. CRC Press, Boca Raton (2004)

[19] McCullough, J., Torng, E.: SRPT optimally utilizes faster machines to minimize flow time. In: Munro, J.I. (ed.) Proc. 15th Symp. on Discrete Algorithms, pp. 350–358. SIAM, Philadelphia (2004)

[20] Pelc, A.: Broadcasting in radio networks. In: Handbook of Wireless Networks and Mobile Computing, pp. 509–528. Wiley and Sons, Chichester (2002)

[21] Schmid, S., Wattenhofer, R.: Algorithmic models for sensor networks. In: Proc. 20th Int. Parallel and Distributed Processing Symposium. IEEE, Los Alamitos (2006)

[22] Schrage, L.: A proof of the optimality of the shortest remaining processing time discipline. Operations Research 16(3), 687–690 (1968)

Target Counting under Minimal Sensing: Complexity and Approximations*

Sorabh Gandhi, Rajesh Kumar, and Subhash Suri

Department of Computer Science,
University of California,
Santa Barbara, CA-93106

Abstract. We consider the problem of counting a set of discrete point targets using a network of sensors under a minimalistic model. Each sensor outputs a single integer, the *number* of distinct targets in its range, but targets are otherwise indistinguishable to sensors: no angles, distances, coordinates, or other target-identifying measurements are available. This minimalistic model serves to explore the fundamental performance limits of low-cost sensors for such surveillance tasks as estimating the number of people, vehicles or ships in a field of interest to first degree of approximation, to be followed by more expensive sensing and localization if needed. This simple abstract setting allows us to explore the intrinsic complexity of a fundamental problem, and derive rigorous worst-case performance bounds. We show that even in the 1-dimensional setting (for instance, sensors counting vehicles on a road), the problem is non-trivial: target count can be estimated within relative accuracy of factor $\sqrt{2}$ and this is the best possible in the worst-case. We then address additional questions related to constructing *feasible* target placements, and noisy counters. In two dimensions, the problem is considerably more complicated: a constant-factor approximation is impossible. Our algorithms and analysis can easily handle some of the non-idealities of real sensors, such as asymmetric ranges and non-exact target counts.

1   Introduction

Inexpensive smart sensors coupled with ad hoc wireless networking provide a compelling and cost-effective technology for what is variously called ubiquitous computing or situational awareness. Specifically, there has been a growing interest in the *networked* power of many cheap and low-fidelity but unattended and geographically-distributed sensors. Because of their low cost, both in hardware that can be several orders of magnitude cheaper than their "mainframe" counterparts, and the untethered, self-organizing architecture that makes them attractive for deployment at large geographic scale without costly human management, pervasive sensor networks hold great potential for "environmental monitoring." The hardware costs and availability, however, are only part of the solution. In order to

---

* This research was supported in part by the National Science Foundation under grants CNS-0626954 and CCR-0514738.

S. Fekete (Ed.): ALGOSENSORS 2008, LNCS 5389, pp. 30–42, 2008.

realize the full potential of these networked smart sensors, significant challenges in algorithms, software, and signal processing must be addressed, many of which arise from the "minimalistic" nature of this sensing and computing platform.

In this paper, we examine some of these key issues in the context of *counting and localizing* targets in a physical space under minimal sensing assumptions. We focus on target counting, as opposed to the more-widely studied target tracking problem, for two reasons: (1) counting is an important problem in its own right; in many environmental monitoring and unattended surveillance applications, for which sensor networks are an ideal platform, accurately estimating a population (e.g. animals in natural habitats, intruders in sensitive areas) is a fundamental end goal; and (2) a good estimate on the target count is often a pre-requisite for robust tracking; for instance, many popular tracking heuristics such as those based on particle filters need a good educated guess on the number of unknown targets to avoid getting stuck.

We frame our research within a minimalistic sensing model to align it with the primary motivation behind the appeal of sensor networks: *low cost and small form factor*. As a result, the binary sensing model has received a great deal of attention for target tracking and other monitoring applications, both in theory and practice (for instance, see [1,2,3,4,5,6]). While the binary sensing model has been shown to achieve excellent performance for tracking a single target [5], for multiple targets it is useful only in settings where the targets are pairwise widely-separated, as was formalized in [6]. As a result, provable-quality tracking and counting of targets requires a richer class of sensors.

In this paper, we work with an abstract model of a *counting sensor*: each sensor outputs an integer value, representing the number of distinct targets in its sensing range. Each target is modeled as a point. The sensor produces no other information about the targets, such as their locations, angles, distances, or any other distinguishing identifiers. While a convenient abstraction for our theoretical investigation of the fundamental limits of target counting and localization, such a sensor is also a fairly good first-order approximation of low-cost radar sensors that can detect the presence of multiple targets but cannot localize them individually. Other sensors including infra-red sensors or acoustic sensors also exhibit this characteristic. In low-cost camera systems as well, achieving reliable calibration or coordinating multiple snapshots for depth and location is both difficult and error-prone [7,8,9]. Furthermore, the measurements are often so noisy that systems actually improve performance by using only the simplest and most robust information content; for instance, Oh et al. [10] report that the variability in the signal strength of their PIR (passive infrared) motion sensors was so great that they actually *improved* the performance of their tracking system by using them as *binary* sensors.

Because our main focus is fundamental achievable limits of performance, we begin with an idealized sensing model, and then discuss the impact of these assumptions as well as generalizations to non-idealized settings. We assume that each ideal sensor has a circular sensing range of a known radius, and it

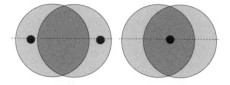

Fig. 1. The two scenarios have identical sensory information: each sensor detects 1 target, yet the total number of targets is different in the two cases

reliably counts the number of distinct targets in its range. Even with such idealization, it is easy to see that our minimal sensing model does not have enough information to accurately count targets *even in 1-dimension*. Figure 1 shows an example of two scenarios with two sensors. The sensory information of both sensors is identical in the two scenarios: both sensors detect 1 target. Thus, there is no way to distinguish between the two scenarios, and decide whether the true target count is 2 (left) or 1 (right). One can, of course, generalize this to an example where sensors cannot distinguish between n and $2n$ targets, and arrive at the impossibility result that, under our minimal sensing model, *no algorithm* can count targets with an *accuracy factor* better than $\sqrt{2}$. It turns out, however, that this is essentially the worst-possible scenario, and one can always achieve $\sqrt{2}$ approximation factor for *any configuration of targets and sensors in one dimension*.

Given our sensing model, one may feel that the best counting accuracy is achieved by non-overlapping sensing ranges—the inaccuracies arise only from multiple sensors counting the same target. Why not just deploy sensors with non-overlapping ranges and obtain the best possible results? There are at least three reasons for sensors with overlapping ranges. First, circles do not tile the two-dimensional plane, and so even in an idealized setting, one cannot achieve full coverage without overlapping circular ranges. Second, while the target count can be improved by *minimizing* the overlap among different sensing ranges, the *location accuracy*, in fact, improves with increasing the overlap [5]. Thus, there is an inherent tension between counting accuracy and the localization accuracy, which may promote sensor deployments with significantly overlapping ranges, even in one-dimensional situations, like a road environment. Finally, all of our results, in fact, hold even when the sensing ranges are not ideal disks; they just need to be connected intervals in one-dimension and any reasonable convex shape in two dimension. Thus, our theory applies to irregular, anisotropic sensing ranges of real sensors, whose overlap is both unpredictable and impossible to eliminate. Therefore, in this work we approach the problem with a worst-case viewpoint, and make no assumptions about the placement of targets or the sensors. We seek to provide worst-case guarantees for the target count for any (adversarial) choice of targets and sensor ranges.

Our approximate counting algorithm, however, is non-constructive, in that it does not necessarily produce a configuration of targets consistent with the sensing input—*it just produces upper and lower bounds on the target population*. Furthermore, it is easy to show examples where not all target counts between the lower and upper bounds are feasible, meaning that there is no possible configuration of targets that is consistent with the sensors' readings. Constructing a feasible configuration of targets is not entirely trivial, but it can be solved in polynomial time by a reduction to the shortest path problem in a graph.

Next, we consider the impact of some non-idealities on our results. In particular, we allow sensor ranges to be non-unit-disk: they can be arbitrary size segments in 1D and arbitrary convex regions in the plane, and they can be asymmetric around the sensor. The target sensing also can be "noisy," in that the number of targets detected by a sensor can lie in an uncertainty range. Specifically, we assume that if the true reading of a sensor is c, then a sensor can report any value in the range $[(1 - \rho)c, (1 + \rho)c]$, where ρ is the *noise* or *uncertainty* parameter, reflecting the false positives and negatives in the sensor's reading. It turns out that all our algorithms and theorems hold even in these more general and realistic models; of course, the accuracy of the target counting now depends on the parameter ρ.

We then consider the target counting problem in two-dimensions and prove that, in the worst-case, no fixed approximation is achievable. An easy $\sqrt{m}$ approximation is possible if the maximum degree of overlap among sensor ranges is m. (This is in contrast to the 1-dimension, where the approximation factor does not depend on the degree of sensing overlap.) All of these results extend to the "noisy" sensor model. All the theorems in this pre-proceedings version are without proofs, the proofs will be included in the conference proceedings.

2   The Counting Sensor Model

We begin with an idealized model of sensing. Each target is modeled as a point, and each sensor is assumed to have a unit-disk sensing range, with perfect sensing: each sensor is able to count precisely the number of targets present in its range. Neither of these assumptions are critical to our algorithms and analysis, as we later discuss, but provide a convenient framework to understand the fundamental limits of target counting. Because the communication requirements of our collaborative counting are so minimal (each sensor only needs to communicate its reading), we abstract away all networking issues in our discussion. In particular, we assume that all the processing occurs at a base station, or a tracker node, that knows the precise geometry of the sensors' locations and ranges. We make no assumptions about the geographic distribution of sensors or targets: *our results are worst-case.*

Throughout, we assume that the targets have fixed locations, and sensors' readings represent a snapshot of the target locations. This view is valuable even in tracking applications when no a priori information is available about the motion of the targets and where the targets can be deliberately evasive, creating an adversarial situation. In such settings, a tracking algorithm is forced to interpolate the motion across snapshots, and therefore must solve the target counting and localization problem considered here.

We begin our discussion by considering the problem in a one-dimensional setting. We imagine targets as points arranged on a line, and a collection of sensors, each with a unit-interval sensing range. It turns out that the exact counting of targets is non-trivial even in this simple setting, and leads to some interesting results. The 1-dimensional setting is also a useful framework in many practical situations, such as counting targets along a road or counting objects in a crowd using far away cameras.

3 Counting and Localization Targets in One Dimension with Ideal Sensors

We begin by repeating our earlier example to argue that precise counting is not possible even in one dimension, and even with idealized counting sensors.

Theorem 1. *If sensors have overlapping ranges, then precise counting of targets is impossible even with idealized counting sensors. Thus, for arbitrary arrangements of sensors and targets, no algorithm can determine the target count precisely.*

Fortunately, it turns out that this is the worst possible scenario, and the $\sqrt{2}$ approximation of the target count is possible for any (adversarial) placement of targets and sensors in 1-dimension.

3.1 Target Count Approximation

Let $S = \{s_1, s_2, \ldots, s_n\}$ denote the set of sensors, and let $C = \{c_1, c_2, \ldots, c_n\}$ denote their sensing counts; that is, c_i is the number of targets detected by s_i in its range. We denote the set of sensing ranges by R, and the union of all these ranges by U. Recall that each sensing range is an interval on the line containing the sensors and the targets. We assume that U is a contiguous range, if not, we run our algorithm on the disconnected contiguous subsets of U separately and add the counts to get the approximate count.

Our algorithm for approximating the number of targets, which we call the SCAN algorithm, is as follows. We compute a *non-redundant* subset $R' \subset R$ of the sensing ranges, where non-redundancy means that union of the ranges in R' equals U, and no range $r \in R'$ is covered by the union of the remaining ranges in the set. In other words, no range can be deleted from R' without losing some coverage of the domain.

Let us denote the set of sensors associated with R' by $S' = \{s'_1, s'_2, \ldots, s'_{n'}\}$ and their readings by set $C' = \{c'_1, c'_2, \ldots, c'_{n'}\}$. Our algorithm outputs $C_A = \frac{S_{C'}}{\sqrt{2}}$ as the target count, where $S_{C'}$ is the sum of readings of the set C', namely, $S_{C'} = \sum_{1 \leq i \leq n'} c'_i$. The algorithm for finding the set R' is given in Algorithm 1.

It is easy to verify that this algorithm can be implemented in worst-case time $O(n \log n)$. We now prove the main result of this section that C_A is a factor $\sqrt{2}$ approximation of the true count, which we denote as C_{OPT}.

Algorithm 1. SCAN

1: Sort the segments in R in increasing order of left endpoints.
2: $R' = \emptyset$, $r =$ first segment in the sorted set R
3: $R' = R' \cup \{r\}$, $R = R \setminus \{r\}$
4: **while** $R \neq \emptyset$ **do**
5:    $T =$ set of segments in R that intersect with r.
6:    Let r' be the segment in T with the rightmost endpoint.
7:    $R = R \setminus T$, $R' = R' \cup \{r'\}$, $r = r'$.
8: **end while**
9: Output the total count of targets associated with ranges in R divided by $\sqrt{2}$.

Theorem 2

$$\frac{C_A}{\sqrt{2}} \leq C_{OPT} \leq \sqrt{2}C_A$$

In effect, the algorithm SCAN above outputs a range $[a, b]$ with a guarantee that the true target count lies between a and b, and $b \leq 2a$. By predicting the geometric mean of these two bounds as an approximation, the algorithm can guarantee that its prediction is within a factor of $\sqrt{2}$ of the true count.

Unfortunately, the counting scheme presented so far is non-constructive—it tells us bounds on the number of targets, but offers no actual placements of targets satisfying the readings of all the sensors. In the following section, we address this fundamental problem of producing target placements consistent with the sensors' readings.

3.2   Target Placement

Consider the example in Figure 2. For this example, the algorithm SCAN outputs the target range $[2, 4]$, which clearly is consistent with the sensors' readings. However, a moment's reflection shows that there is no realizable (feasible) *target placement* that is consistent with the target count of either 2 or 3. Indeed, the only feasible target placement satisfying the sensors' readings needs 4 targets, as shown. Even for feasible target counts, the algorithm does not provide an actual placement of targets. We address these shortcomings in the following.

Consider a set $S = \{s_1, s_2, \ldots, s_n\}$ of n sensors along the X-axis, and let $C = \{c_1, c_2, \ldots, c_n\}$ denote the readings associated with these sensors. Let $P =$

Fig. 2. An example of 3 sensors on a line, where the first and the third sensor has target count 2, while the middle sensor has count 0. The SCAN algorithm outputs a target range of $[2, 4]$. Only the target count of 4 is realizable as a physical configuration of targets consistent with the sensors' readings.

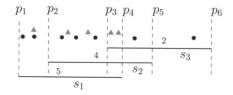

Fig. 3. An example with 7 targets and 3 sensors. The true target positions are shown as solid circles. The sensor readings are shown by the numbers placed above each sensor's range. The output placement computed by our algorithm is shown using lightly shaded triangles.

$\{p_1, p_2, \ldots, p_{2n}\}$ denote the set of $2n$ points defining the start and the end points of the sensor ranges, sorted in order of increasing x-coordinates; that is, the x-coordinate of p_i is less than the x-coordinate of p_{i+1}.

We introduce a set of variables $Z = \{z_1, z_2, \ldots, z_{2n}\}$ where z_i represents the *total* number of targets lying to the left of point p_i. By definition, therefore, we have the following constraint:

$$z_1 \leq z_2 \leq \ldots \leq z_{2n}, \tag{1}$$

because the number of targets to the left of p_{i+1} is at least as large as the number of targets to the left of p_i.

Next, if p_j and p_k are the starting and end points associated with the range of sensor s_i, then $z_k - z_j$ denotes the number of targets in s_i's range. This introduces another constraint:

$$z_k - z_j = c_i \tag{2}$$

We have one such constraint for each sensor. Any assignment of z_i's satisfying these constraints, together with $z_1 = 0$, corresponds to a feasible placement of targets for our problem. In particular, a feasible solution can be obtained by placing $z_i - z_{i-1}$ targets spaced equally between points p_{i-1} and p_i, for $2 \leq i \leq 2n$. The set of constraints described above can be solved as an integer linear program. Unfortunately, in general, integer linear programming is NP-Hard. Fortunately, the special structure of our problem admits a rather efficient (polynomial time) solution, by a transformation to a shortest path problem. In particular, all the constraints in our problem have the form of a *difference constraint*. We explain the reduction to the shortest path problem, using an example.

Consider the example shown in Figure 3, with 7 targets and 3 sensors. The true target positions are shown as solid circles. The sensor readings are shown by the numbers placed above each sensor's range. The first set of constraints that enforce the conditions $z_i \leq z_{i+1}$, for $1 \leq i \leq 2n$, can be written as the following set of difference constraints:

$$z_1 - z_2 \leq 0, \quad z_2 - z_3 \leq 0, \quad z_3 - z_4 \leq 0$$
$$z_4 - z_5 \leq 0, \quad z_5 - z_6 \leq 0$$

Each of the equality constraint encoding the count of each sensor (Eq. 2) can be written as a pair of difference constraints:

$$z_4 - z_1 \leq 5, \quad z_1 - z_4 \leq -5$$
$$z_5 - z_2 \leq 4, \quad z_2 - z_5 \leq -4$$
$$z_6 - z_3 \leq 2, \quad z_3 - z_6 \leq -2$$

These inequalities can be transformed into the formulation of a shortest path problem in a graph as shown in Figure 4. In this graph, there exists a node for each variable z_i, and an edge for each difference constraint. In particular, the difference constraint $z_i - z_j \leq \ell$ maps to an edge directed from node w_j to node w_i, with weight ℓ. In addition, we add an artificial node s, and introduce 0-weight edges from s to all other nodes in the graph. We now observe that this graph has well-defined shortest paths from s to all other nodes *if and only if* there is no negative-weight cycle in the graph. More precisely, if there is a negative cycle in the graph, then the set of inequalities are inconsistent, and there is no feasible solution. Otherwise, the shortest path distances to the nodes z_i correspond to a feasible solution.

Solving the shortest path problem on the graph gives the following shortest path distances from s: $z_1 = -5$, $z_2 = -4$, $z_3 = -2$, $z_4 = z_5 = z_5 = 0$. We can enforce $z_1 = 0$ by adding 5 to all these variables, without violating any constraints. We then get: $z_1 = 0$, $z_2 = 1$, $z_3 = 3$, $z_4 = z_5 = z_6 = 5$. The placement of targets corresponding to these variable settings is shown in Figure 3 by lightly shaded triangles.

We can solve the shortest paths problem in the graph using the Bellman-Ford algorithm; this algorithm either determines that the graph contains a negative-weight cycles, or computes valid shortest path distances in worst-case time $O(|V||E|)$, where $|V|$ and $|E|$, respectively, are the number of vertices and edges in the graph [11] . In our setting, both the number of vertices and edges is $O(n)$, so the algorithm has time complexity $O(n^2)$. We can now state the main result of this section.

Theorem 3. *Given a set of n counting sensors on a line and their target counts, we can find in $O(n^2)$ time a placement of targets consistent with all the sensors' counts, or determine that the sensors' readings are inconsistent.*

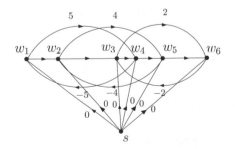

Fig. 4. The graph corresponding to the example of Figure 3

The Bellman-Ford shortest path algorithm, in fact, has an interesting property: the algorithm minimizes the maximum difference between the difference variables. In other words, the algorithm not only finds a feasible assignment of variables, but actually finds an assignment satisfying

$$\min \quad \max_{1 \leq i \leq 2n, 1 \leq j \leq 2n} |z_i - z_j|$$

In our setting, the maximum difference is between the variables z_{2n} and z_1 (Equation 1). But $z_{2n} - z_1$ equals the total number of targets in the solution, and so our algorithm finds a feasible solution with the least possible number of targets consistent with the sensors' counts.

4   Extensions to Non-ideal Sensing

In this section, we make a limited attempt to address two of the most severe assumptions of the idealized sensor model, namely, the *unit disk* sensing range and *perfect target count*. In particular, we show that our algorithms can easily handle sensing ranges that are neither unit length (in 1d) or unit disk (in 2d) nor symmetric about the center. Secondly, our algorithms can gracefully handle *noisy* target counts by sensors. Specifically, if the true target count for a sensor is c, then a sensor can report any value in the range $[(1 - \rho)c, (1 + \rho)c]$, where ρ is the noise parameter, reflecting the false positives and negatives in the sensor's reading. We now discuss the implications of these non-idealities on our algorithms.

4.1   Target Count Approximation with Non-ideal Sensors

Let $S = \{s_1, s_2, \ldots, s_n\}$ denote the set of sensors. We denote the set of sensing ranges by R. As in the case of ideal sensors, we use the algorithm SCAN to compute the non-redundant set R'. Let $S' = \{s'_1, s'_2, \ldots, s'_{n'}\}$ denote the set of sensors associated with R' and let $C' = \{c'_1, c'_2, \ldots, c'_{n'}\}$ denote the (noisy) counts associated with these sensors. Our algorithm outputs $C_A = \dfrac{S_{C'}}{\sqrt{2(1-\rho^2)}}$ as the target count, where $S_{C'}$ is the sum of readings of set C', namely, $S_{C'} = \sum_{1 \leq i \leq n'} c'_i$. Let C_{OPT} denote the actual count of the number of targets in the system. The following theorem analyzes the accuracy of this approximation.

Theorem 4

$$\frac{C_A}{\sqrt{\frac{2(1+\rho)}{1-\rho}}} \leq C_{OPT} \leq \sqrt{\frac{2(1+\rho)}{1-\rho}} C_A$$

It is not too difficult to see that our bounds for both ideal and non-ideal sensors are the *best possible* in the worst-case. In particular, given any value of ρ, it is possible to achieve the worst-case approximation factor (both overcount and undercount) with just two sensors. In the next section, we extend the target placement algorithm proposed for ideal sensors to estimate target placements in the presence of non-ideal sensors.

4.2 Target Placement with Non-ideal Sensing

Let the sets $C' = \{c'_1, c'_2, \ldots, c'_n\}$ and $C = \{c_1, c_2, \ldots, c_n\}$ denote the noisy and the true readings of the sensor set $S = \{s_1, s_2, \ldots, s_n\}$. Let us associate the set of variables $Z = \{z_1, z_2, \ldots, z_{2n}\}$ with the sorted set $P = \{p_1, p_2, \ldots, p_{2n}\}$ of start and end point of the sensor ranges, where, as for ideal sensors, z_i denotes the number of points to the left of point p_i. We now show that a feasible target placement can be obtained even for non-ideal sensors,

Theorem 5. *Given a set of n non-ideal sensors and their readings, we can find a placement for targets in the network which satisfies all sensor readings.*

Of course, both the target placement as well as the number of targets estimated may be different from the ones found using non-noisy counts, but the approximations are guaranteed to be within the range of accuracy given by our theorems.

In the next section, we consider the target counting problem when the sensors and the targets are scattered in a two-dimensional plane.

5 Two Dimensional Target Counting

We begin with an example to argue that, unlike in the one dimension, approximation within a constant factor is not achievable for the two-dimensional target counting problem. The construction is quite simple, and shown in Figure 5. Imagine starting with n circles, centered at the origin. (The circles represent the sensing ranges of our idealized counting sensors.) We keep one circle stationary, and translate the centers of the remaining $n-1$ circles by $\{\delta, 2\delta, \ldots, (n'-1)\delta\}$ along the positive X-axis, where δ is chosen such that $(n-1)\delta$ is less that the radii of these circles.

With this arrangement of sensors, consider two different sets of target placements. In the first case (left figure), we place k targets at the origin. In the second case (right figure). we place k targets each near the top of each sensor's range. It is easy to see that in both cases, each sensor counts precisely k targets in its range, but the total number of targets present is k in the first case, and nk in the second case. Because the two cases are indistinguishable based on the

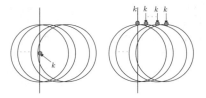

Fig. 5. Two scenarios with identical sensors information, but different total target counts: k for the left figure, and nk for the right figure

sensors' counts, no algorithm in this model can count targets with any constant factor accuracy. We summarize this result in the following theorem.

Theorem 6. *For arbitrary arrangements of sensors and targets, no algorithm can achieve a constant-factor approximation of the target counts even with idealized counting sensors in two dimensions.*

Clearly, the main source of difficulty is the overlapping sensing ranges. The construction of Figure 5 achieves the impossibility result by forcing an unbounded level of overlap among the ranges. In the following, we argue that if the degree of overlap is at most m, then one can approximate the target count within factor $\sqrt{m}$; thus, in practice, where we expect the overlap to be small, the approximation may be acceptable. By the degree of overlap, we mean the maximum number of sensing ranges that cover a point in the plane.

Theorem 7. *If the maximum degree of sensor overlap is m, then we can approximate the total number of targets in two dimensions to within a factor $\sqrt{m}$.*

If we consider the non-idealities of convex ranges and noise parameter as defined in Section 2, then for these non-ideal sensors the proof given above can be extended to obtain an approximation factor of $\sqrt{\frac{m(1+\rho)}{1-\rho}}$.

6   Related Work

The problem of detecting and tracking targets is of broad interest to many applications dealing with unattended monitoring and surveillance, with a rich literature in many disciplines, including computer vision, signal processing, ad hoc networks etc. [7,12,9,13]. The research goals in these areas, however, are different from those being pursued in sensor networks. In particular, the vision and signal processing communities are concerned with extraction of distinguishing features in detailed signals (e.g. images) and classifying the targets (e.g. tanks or cars). The mobile and ad hoc network communities have considered tracking with the goal of maintaining the state of network connectivity. In these cases, the nodes try to track other nodes using mobility models so that routing can be achieved successfully.

Counting targets is closely related to monitoring, intrusion detection and tracking targets. Counting is often the first step in most of these applications. Research in sensor networks has seen a lot of work in tracking multiple targets [14,15,16,17,18,19,20,21] and almost every piece of work assumes that the number of targets in the network is known. Our work is closely related to [5,6] in terms of deriving fundamental limits for tracking and counting targets using a minimal sensing model. These papers use a binary sensing model, which has also been considered by [1,2,3,4]. Counting sensor model is similar to the binary model in terms of minimal sensing, instead of transmitting a bit of information, counting sensors transmit an integer representing the number of targets in their range. The problem of counting targets is explored in [6] and the authors show that even in one dimension, counting targets accurately is not possible using binary sensing model unless the targets are spaced far apart from each other. Gfeller et. al. [22] add to the basic binary sensing model by considering mobile binary sensors and show that even then the problem is hard. Counting targets is also addressed in [23], the sensors considered are proximity sensors and sense the amplitude only. The target counting is then represented as peak counting problem in the aggregate sensor network, but the framework assumes that the targets are well separated. In [24], the authors look at the problem of counting the number of people in a crowd using image sensors. They subtract background from the image and then count number of visual hulls to count number of people. Their focus is on geometric hull computation and our techniques can be used on top of their algorithms to provide the bounds and counts. In [25] the authors use topological integration theory to provide expected target counts as compared to the deterministic bounds provided in this paper. The model of sensing considered is similar in the sense that the sensors give a count of the number of objects. However, the paper does not make any assumptions about sensing shapes and proves that the expected counts is the best one can hope for without geometry.

References

1. Arora, A., Dutta, P., Bapat, S., Kulathumani, V., et al.: A line in the sand: A wireless sensor network for target detection, classification, and tracking. Computer Networks (2004)
2. Aslam, J., Butler, Z., Constantin, F., Crespi, V., et al.: Tracking a moving object with a binary sensor network. In: SENSYS (2003)
3. Lazos, L., Poovendran, R., Ritcey, J.: Probabilistic detection of mobile targets in hctcrogeneous sensor networks. In: IPSN (2007)
4. Oh, S., Sastry, S.: Tracking on a graph. In: IPSN (2005)
5. Shrivastava, N., Mudumbai, R., Madhow, U., Suri, S.: Target tracking with binary proximity sensors: fundamental limits, minimal descriptions, and algorithms. In: SENSYS (2006)
6. Singh, J., Kumar, R., Madhow, U., Suri, S., et al.: Tracking multiple targets using binary proximity sensors. In: IPSN (2007)
7. Cai, Q., Aggarwal, J.K.: Tracking human motion using multiple cameras. In: ICPR (1996)
8. Nguyen, N., Venkatesh, S., West, G., Bui, H.: Multiple camera coordination in a surveillance system (2003)
9. Stauffer, C., Eric, W., Grimson, L.: Learning patterns of activity using real-time tracking. IEEE Transaction on Pattern Analysis Machine Intelligence (2000)
10. Oh, S., Chen, P., Manzo, M., Sastry, S.: Instrumenting wireless sensor networks for real-time surveillance. In: International Conference on Robotics and Automation (2006)
11. Cormen, T., Lieserson, C., Rivest, R., Stein, C.: Introduction to algorithms (2004)
12. Collins, R., Lipton, A., Fujiyoshi, H., Kanade, T.: Algorithms for cooperative multisensor surveillance. Proceedings of the IEEE 89, 1456–1477 (2001)
13. Zaidi, Z., Mark, B.: A mobility tracking model for wireless ad hoc networks. In: IEEE WCNC (2003)

14. Chen, H., Kirubarajan, T., Bar-Shalom, Y.: Multiple target tracking with multiple finite resolution. In: 5th International Conference on Information Fusion (2002)
15. Hwang, I., Roy, K., Balakrishnan, H., Tomlin, C.: A distributed multiple-target identity management algorithm in sensor networks. In: IEEE Conference on Decision and Control (2004)
16. Jung, B., Sukhatme, G.: Tracking targets using multiple robots: The effect of environment occlusion. Autonomous Robots (2002)
17. Liu, J., Liu, J., Reich, J., Cheung, P., et al.: Distributed group management for track initiation and maintenance in target localization applications. In: Zhao, F., Guibas, L.J. (eds.) IPSN 2003. LNCS, vol. 2634, pp. 113–128. Springer, Heidelberg (2003)
18. Mechitov, K., Sundresh, S.: Cooperative tracking with binary-detection sensor networks. In: SENSYS (2003)
19. Oh, S., Hwang, I., Roy, K., Sastry, S.: A fully automated distributed multiple-target tracking and identity management algorithm. In: AIAA Guidance, Navigation, and Control Conference (2005)
20. Rachlin, Y., Negi, R., Khosla, P.: Sensing capacity for discrete sensor network applications. In: IPSN (2005)
21. Shin, J., Guibas, L., Zhao, F.: A distributed algorithm for managing multi-target identities in wireless ad-hoc sensor networks. In: Zhao, F., Guibas, L.J. (eds.) IPSN 2003. LNCS, vol. 2634, pp. 223–238. Springer, Heidelberg (2003)
22. Gfeller, B., Mihalak, M., Suri, S., Vicari, E., et al.: Counting targets with mobile sensors in an unknown environment. In: Kutyłowski, M., Cichoń, J., Kubiak, P. (eds.) ALGOSENSORS 2007. LNCS, vol. 4837, pp. 32–45. Springer, Heidelberg (2008)
23. Fang, Q., Zhao, F., Guibas, L.: Counting targets: Building and managing aggregates in wireless sensor networks. PARC Technical Report (2002)
24. Yang, D., Gonzalez-Banos, H., Guibas, L.: Counting people in crowds with a real-time network of image sensors. In: International Conference on Computer Vision (2003)
25. Baryshnikov, Y., Ghirst, R.: Target enumeration in sensor networks via integration with respect to euler characteristic (2007)

Efficient Scheduling of Data-Harvesting Trees

Bastian Katz*, Steffen Mecke**, and Dorothea Wagner

Universität Karlsruhe (TH)
{katz,mecke,wagner}@ira.uka.de

Abstract. Many applications in sensor networks demand for energy and time optimal routing of data towards a sink. In this work we present mechanisms to set up energy and time efficient TDMA schedules for a given routing tree under very strict limitations: Nodes have only a constant size memory and must agree on a schedule using only a minimum of communication for set up: Each node is only allowed to send a single message to each of its neighbors.

We propose and analyze solutions in two different interference models. We show that, despite these tight restrictions, it is possible to compute energy optimal schedules which are almost time optimal and time optimal schedules which are almost energy optimal in the total interference model and we describe a 4-approximative algorithm in the k-local interference model.

We also show how to extend these mechanisms to settings with packet loss, while still guaranteeing bounds on energy consumption.

1 Introduction

Data-harvesting applications in sensor networks gather bulk data from the data field and collect them at a central repository or sink. Examples are data archiving or surveillance applications that periodically sample snapshots at high rates, storing or analyzing them centrally outside the network [1]. Another class are scenarios in which network nodes store measured data until from time to time a user requires access to the data stored in a large number of nodes [2]. In low-power networks, these applications demand highly optimized communication management to keep the network operable for as long as possible.

This paper considers sensor networks where sensor nodes are distributed over a geographic area and measure values in regular time intervals. At certain times, the stored data must be routed through the network and collected at a central location, the *sink*, usually along a routing tree rooted at the sink. Since the radio communication dominates the energy consumption, minimizing the cost of wireless communication is crucial to maximize the lifetime of a sensor network.

* Partially supported by the German Research Foundation (DFG) within the Research Training Group GRK 1194 "Self-organizing Sensor-Actuator Networks".
** Partially supported by the "BW-FIT" project "ZeuS" by the Landesstiftung Baden-Württemberg.

S. Fekete (Ed.): ALGOSENSORS 2008, LNCS 5389, pp. 43–56, 2008.

Power consumption in this scenario can be reduced in several ways. First by compressing the data, second by improving the routing tree to prevent that single nodes are overly burdened, and third by avoiding all unnecessary power consumption of the radio. Although these three issues are not fully independent, it makes sense to analyze them separately. Data reduction is an application-specific problem that is largely orthogonal to the other two problems. The construction of routing trees is subject to many other practical restrictions such as link quality in real-life networks. We will thus focus on the problem of avoiding all unnecessary power consumption by the radio for a fixed routing tree provided by some arbitrary protocol.

If communication patterns are known in advance, schedule-based ("TDMA") protocols outperform contention-based ("CSMA") protocols, because they do not waste energy due to idle listening and collisions. Their drawback however is an increased protocol overhead for schedule set-up. Therefore, we develop and analyze efficient, schedule-based protocols that require only a minimum of set-up communication.

The problem addressed in this paper can be summarized as follows: Given a routing tree in which all nodes store data that is to be collected at a sink, we allow each node to pass only one packet to each of its neighbors in that tree. Is it possible to agree on an energy-optimal TDMA schedule, i. e., a schedule that allows to transport all stored packets to the sink and in which all nodes know exactly when to send and when to listen? Is it possible to agree on a schedule of minimum length?

We will contribute to this problem by providing lower bounds and algorithms for two reasonable interference models. In the total interference model, we assume that at any time, only one single transmission is allowed throughout the whole network. We will provide two transmission schedules that comply with the above restrictions, the first being time-optimal at the price of adding a small protocol overhead to routed packets, the second being energy-optimal, but not time-optimal. We also provide a third transmission schedule for routing trees that is both, energy- and time-optimal, at increased set-up costs. Furthermore, we prove that in this interference model, an energy- and time-optimal transmission schedule cannot route single data packets with minimum delay. We conjecture that it is impossible to set up an energy- and time-optimal schedule under our restrictions at all.

In the k-local interference model we assume that transmissions do only interfere with transmissions within some constant neighborhood. We will show how to set up a transmission schedule that is energy-optimal and constant-factor time-approximative for the given tree.

This paper is organized as follows: In the remainder of this section, we will discuss related work and give a formal problem definition. In Section 2, we cover the problem of finding schedules for a routing tree with total interference. In Section 3, we do the same for local interference models. Section 4 discusses a method to handle the problem of unreliable links. We conclude in Section 5.

1.1 Related Work and Overview

There is a plethora of algorithms for finding topologies (or routing information) for the data gathering problem in sensor networks. There are several goals for optimization, for example throughput, latency, reliability, security and energy consumption, the last one being the most important in sensor networks. Almost all of these different approaches, however, construct one or sometimes several routing *trees*.

There has been previous work on minimzing the *time* for data gathering. In [3] a problem similar to ours is studied, a 4-approximation algorithm is given and NP-hardness of the problem shown. A problem with variable release times is studied in [4]. Unlike this previous work, however, we focus on distributed algorithms and take set-up cost into account. Also, we do not concentrate so much on time optimality but on *energy* optimality.

There are two main kinds of medium access methods: contention based ("CSMA") protocols and scheduled ("TDMA") protocols (see [5] for a partial overview of MAC protocols for sensor networks). There are also a few hybrid forms. The strength of contention based protocols include simplicity, low protocol overhead and flexibility, but they suffer from energy waste caused by collisions, overhearing and idle listening. TDMA-based protocols do not have any of these drawbacks (at least in theory) but they require more communication to establish the scheme, time synchronization is usually more of an issue and they are less flexible in case of topology changes.

One of the very few contention-based MAC protocols that take advantage of the tree topologies present in data-archiving systems is [6]. The wakeup times of nodes are staggered on paths towards the sink, which reduces latency. Measures are employed to reduce interference among packets travelling along the same paths. Additionally, special "More-to-Send" packets are used to further synchronize wakeup times and thus increase throughput. However, this protocol is designed for very low data rates. It is not energy-optimal and there are no special mechanisms to reduce congestion and ensure fairness.

Flexible Power Scheduling (FPS) is described in [7]. In FPS parents are responsible for assigning time slots to their children. FPS reduces contention but does not guarantee collision-free communication. Therefore, an underlying MAC layer is still required. Fairness among children in different branches is not ensured.

MPS (Multi-Flow Power Scheduling) and HPS (Hybrid Power Scheduling) are enhancements on FPS introduced in [8]. MPS is closely related to our k-layer interference protocol in Section 3, but performance is only evaluated experimentally and there is no theoretical analysis of the protocols. Also, the interference model is not described explicitly.

The authors of [2] address the problem of congestion, fairness and robustness during the transport of high volumes of sampled data. They use the total interference model (cf. Section 2). Their approach is based on a slot distribution scheme. Nodes that have no more packets to send can pass their slots back to their parent. Every second slot that a node receives from its children is passed on to its parent. This aims at distributing slots more fairly: Nodes with high loads get more slots. In [9] further refinement of slot distribution strategies are developed. But even with these refinements, the channel usage is still fairly low and decreases with growing network size. In the same paper, there is another scheme (similar to ours in Section 2) which, however, leads to unlimited buffer size and the control message overhead is not analyzed.

DOZER ([10]) is an approach that tries to solve the problems of medium access, tree construction and scheduling together. The authors employ a local TDMA scheme which reduces requirements on clock synchronization but does not ensure fairness. Collisions are reduced by letting schedules of interfering node pairs "drift apart" through randomization. This approach is designed for scenarios with very low data rates but causes problems when there are higher volumes of data to deliver.

In contrast to these approaches, we focus on detailed theoretical analysis of throughput and protocol overhead. Our approaches are also designed for arbitrarily large networks and high data load.

1.2   Problem Definition and Network Model

Throughout this work, we assume that we are running a sensor network with one node serving as a sink that is connected to some infrastructure or monitor. The task now is to set-up a transmission schedule that collects sensor data from all nodes at the sink without aggregation. This task naturally divides into the following subtasks or *stages*:

Topology stage: decide on a topology to collect data. We assume that the resulting topology is a tree routed at the sink.

Set-up stage: perform the communication necessary to agree on a schedule that guarantees delivery of all sensor data to the sink and complies with an interference model.

Collection stage: run the schedule as long as data is to be collected.

Energy consumption is our primary concern. We want to minimize it during both the set-up stage and the collection stage. Energy use during the collection stage can be minimized if each node knows exactly in which slots to listen and when to send and if there is no idle listening or failed transmission. Nodes that are neither sending nor trying to receive can go into sleep mode or at least turn off their radio, dramatically reducing the energy consumption. We concentrate on solutions which achieve this with a minimalistic kind of set-up communication: Just one convergecast and one broadcast. It is impossible to let all nodes know about the schedule's length, let alone a first point in time to transmit or receive with less communication.

More formally, we restrict solutions to the following model:

1. Each node v in the network must transmit a (possibly individual) number of own data packets, $\sigma(v)$ to the sink.
2. Within the network, a spanning tree T, rooted at the sink r is provided by some standard protocol, i.e., every node knows its parent and a list of its children.

3. During the set-up stage, every node can send at most one packet of size $O(\log N)$ bits to each of its neighbors in T, N being the number of packets in the network. We assume that during this stage a different medium access scheme is used to establish parameters for the TDMA-based scheme of the collection stage.
4. Each node (except for the sink) has only a limited, constant amount of memory for buffering packets and storing information about the slots to be active.

As a secondary criterion, we want to minimize the length of the schedule, i.e., the time until all packets have been delivered.

1.3   Definitions and Notation

We will denote a node's distance from the sink in T by $h(\cdot)$ and the height of the tree by h. We will talk of *children*, *descendants*, *parents* and *ancestors* in the usual sense. We will refer to the set of a node v's descendants including v as $D(v) := \{w \mid w$ is descendant of $v\} \cup \{v\}$. The set of children of v is denoted by $C(v)$. We will assume that there is a fixed ordering among the children of every node. A child v is said to be *left* of w if it preceeds w in this order. We will also refer to the pre- and postorder number of nodes as $\text{pre}(v)$ and $\text{post}(v)$ (with respect to these orderings) in the usual sense. As defined above, let $\sigma(v)$ denote the number of own packets a node has to deliver. We will assume that every node (except the sink, for which we assume $\sigma(r) = 0$) has at least one data packet. We will shortcut $\sum_{v \in V'} \sigma(v)$ by $\sigma(V')$ and $\sigma(V)$ with N. We will also write $\pi(V') := \sum_{v \in V'} h(v)\sigma(v)$ for the number of transmissions a set of nodes causes and $\bar{\pi}(v) := \pi(D(v))$.

2   Scheduling a Tree with Total Interference

This section covers the case of *total interference*. That is, no two nodes are ever allowed to transmit in the same time slot.

2.1   Infeasibility

We conjecture that it is not possible to find a scheme that achieves optimal time *and* energy using only one concast and convergecast in the total interference model under the assumptions given in section 1.2. We have no proof for this claim, but if we restrict ourselves to certain kinds of schedules, it follows quite easily:

Proposition 1. *Let $\sigma(v) \equiv 1$, every node can store at most one packet (at the beginning every node's memory is filled with its own packet), during the set-up stage only one convergecast and one broadcast of (\log N)-sized messages is allowed and no additional information can be attached to the packets during the collection stage.*

Under these restriction it is not possible to compute and run a time-optimal schedule in which each packet is immediately passed to the sink once it has started from its originating node.

Proof. Let r be the sink node. During the set-up stage, r has only received $|C(v)|c\log N$ bits for some constant c. For every packet arriving at r during the collection stage, the height of the originating node can be determined by the time since the arrival of the previous packet. As r is only allowed to be awake when packets arrive, it must know the times of arrival in advance. This is equivalent to knowing the heights $\{h_1, h_2, \ldots, h_n\}$ of the packets in advance. There are, however, an exponential[1] number of such sequences, each one requiring a different schedule at r. But these cannot be identifed by the $C(v)c\log N$ bits.

Most of the assumptions can be relaxed without invalidating Proposition 1, but it is not apparent if different kinds of schedules (which, for example, buffer incoming packets) could achieve more. The rest of the paper will show what is possible if some of the assumptions are relaxed.

2.2 An Optimal Scheme with Increased Packet Size

In our first approach we will achieve a time and energy optimal scheme by allowing slightly increased packet size during the collection stage.

We will proceed as follows: The packets are transported to the sink in preorder. Each packet is immediately passed down to the sink once it has started. A node v never has to wake up before all nodes which are further left than v but not descendants of v have transmitted their packets to the sink. Let's call the first time slot in which v has to transmit a packet $T(v)$. We have

$$T(v) = T(\text{parent}(v)) + \pi\,(\text{parent}(v)) \quad (1)$$

if v is the leftmost child of its parent. Else, we have

$$T(v) = T(s_l(v)) + \bar{\pi}\,(s_l(v)) \ . \quad (2)$$

The $\sigma\,(v)$ packets of v are then transmitted in time slots $T(v), T(v) + h(v), \ldots$ until $T(v) + (\sigma\,(v) - 1)h(v)$. After transmitting its own packets, the next time to wake up for v is when its leftmost child w transmits its packet.

The ancestors of v also have to have information about when to receive these packets on their way to the sink. A node w on the path from v to the sink r has to receive in slot $T(v) + h(v) - h(w) - 1$ and has to send in slot $T(v) + h(v) - h(w)$ (and so on...). Therefore, v attaches

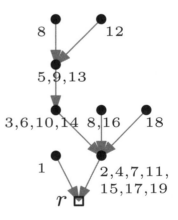

Fig. 1. Example schedule, numbers denote slots for sending

---

[1] Exponential in n, even disregarding isomorphisms and order.

to the payload P of the message an information t about the next time slot it is going to send in.

$$
t = \begin{cases}
T(v) + h(v)k & \text{if } P \text{ is the } k\text{th packet of } v \text{ and it has packets left} \\
T(c) + 1 & \text{if the next packet belongs to a child } c \text{ of } v \\
t' + 1 & \text{if } P \text{ is from a descendant of } v \text{ and } t' \neq 0 \\
0 & \text{if } t' = 0 \text{ and no children of } v \text{ are left}
\end{cases} \tag{3}
$$

After all descendants of v have transmitted their packets, v can sleep for the rest of the protocol.

In summary, each node v has to compute

1. the number of packets $\sigma(D(c))$ in the subtree of each of its children c_i (counted during the convergecast),
2. its own height $h(v)$ (determined during the broadcast),
3. the weighted sizes of the subtrees $\bar{\pi}(c_i)$ (computed from $h(v)$ and information collected during the convergecast),
4. its own starting time slot $T(v)$ and the starting slots of all of its children (computed via Equations (1) and (2)).

Theorem 1. *There is a scheme that produces an optimum length schedule of length $\bar{\pi}(r)$. In this scheme, every node receives $1 + C(v)$ packets and transmits $1 + C(v)$ packets of size $\log(N)$ in the set-up stage.*

Every node receives and sends $\Theta(\sigma(D(v))\log(h))$ additional bits during the collection stage. In total, an additional $\Theta(n\Delta \cdot \log(N) + \bar{\pi}(r)\log h)$ bits are transmitted and received (with $\Delta := \max_v C(v)$). A node needs additional memory of $O(\Delta \log(nh \max_{w \in V} \sigma(w)))$.

Remark 1. The amount of additional data can be reduced to $\Theta(D(v)\log h)$ (or $O(n \log h)$) bits per node. The total amount of additional data can be reduced to $\sum_v h(v) \log h \leq nh \log h$ bits.

Summarizing, we have described a time- and energy-optimal scheme at the price of increasing each packet by $\log(h)$ bits.

2.3 Variants

In the previous section we saw how an optimal (shortest) schedule can be constructed with relatively little (but unbounded) message overhead. With slight modifications (which include adding a memory buffer for one packet at every node), a similar scheme as in the previous section can be constructed for postorder. In this section we will propose two different ways of cheating to get a time optimal schedule.

Our first proposal needs more energy in the set-up stage and more than constant memory:

Proposition 2. *It is sufficient to send (and store) $h \log N$ bits per edge once: the number of nodes in every layer below that edge. If we use level order, we can compute t from this information.*

Finding a scheme for level order seems a little bit more involved. If, however, every node v knows the number of nodes in lower levels and for each level $\ell \geq h(v)$ the number of nodes in level ℓ to the left of v, in v's subtree and to the right of v, then it can easily compute the slots in which it has to send or receive. However, this requires additional memory per node in the order of $\Theta(h \log N)$ and the same amount additional communication *per edge*.

The second approach is a scheme which fulfills all the requirements on communication and memory but at the cost of time optimality:

Proposition 3. *If all nodes have the same number σ of packets, we can arrange an energy optimal schedule with length Nh (h is the tree height) and time approximation factor $\sqrt{n}$.*

Proof. This schedule works as follows: As before we process packets in preorder and immediately hand them down to the sink. The only difference being that $T(v) = \sigma h \cdot (\text{pre}(v) - 1) + h - h(v) + 1$. The packets of v are then transmitted in slots $T(v), T(v) + h, \ldots, T(v) + (\sigma - 1)h$. If v actually has height $h(v) < h$, there are $h - h(v) - 1$ unused slots.

The approximation ratio of the length of this scheme is $\Theta(\sqrt{n})$: For a fixed height h the worst case is a "flower" with one path of length h and all other nodes at height 1. The approximation ratio is $\sigma nh/|S_{\text{OPT}}| = nh/(h \cdot (h-1)/2 + (n-h) \cdot 1) \leq c\sqrt{n}$.

Remark 2. For "well-behaved", geometric graphs in the plane with height $\Theta(\sqrt{n})$ and $O(\ell)$ nodes in every level ℓ (like a regular, geometric grid), the schedule has length $\Theta(\sigma n\sqrt{n})$ which is within a constant factor of the optimum.

3   Scheduling with Local Interference

In this section, we assume that a transmission (u, v) is successful if and only if u is the only active sender in v's k-hop neighborhood for some constant k. We call this interference model k-*local interference*. This model is a reasonable, yet a cautious approximation of interference in dense networks, in which euclidian distance and hop-distance closely correlate.

Quite typically, routing trees for data aggregation in sensor networks are set up using some kind of request flooding, i.e., the routes between a node and the sink are shortest paths (in hops). This is not only a very lightweight, robust protocol, but also guarantees that data travels on short routes, which reduces the risk of packet loss. In such trees, a transmission (u, v) is always successful if u is the only active sender among all nodes with $|h(v) - h(u)| \leq k$. This follows from the triangle inequality. We call this property of a rooted tree k-*layer bounded interference*. We will propose a protocol, coined k-LS, to set up a schedule that is energy optimal and time approximative within a constant factor.

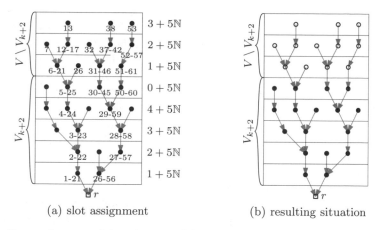

(a) slot assignment	(b) resulting situation

Fig. 2. Slot assignment (a) and result (b) of the first phase of k-LS for $k = 3$ and $\sigma \equiv 1$. Edges are active at time t if t is in the given range and matches the modulo for the sender's level of the tree.

Theorem 2. *In the k-local interference model, there is a scheme for trees with k-layer bounded interference yielding an energy optimal schedule that is time approximative within a factor of $\frac{k+2}{\lfloor (k+1)/2 \rfloor} \leq 4$.*

Proof. The key idea of k-LS is to set up two schedule phases that are performed successively. In the first phase data is "pipelined" towards the sink ending in a state where no nodes having a height of more than $k+2$ have any packets left. In more detail, the sink's neighbors pass packets to the sink every $k+2$ slots starting with the first slot and sending one after another. Whenever a node transmits a packet, it receives a packet in the next slot from one of its children as long as their children have more packets. Every node v passes $\sigma \left(D(v) \setminus V_{k+2} \right)$ packets, i. e., as many packets as there are in its subtree in heights more than $k + 2$. In the second phase data is collected from the remaining nodes using essentially the same technique as in Proposition 2. For the following formal description, we will shortcut the set of nodes with height less than or equal to some h with $V_h := \{u \in V \mid h(u) \leq h\}$.

During the convergecast phase each node v learns how many packets to relay from nodes with heights $h(v) + 1, \ldots, h(v) + k + 2$ and from more distant levels. Obviusly, $(k + 3) \mathrm{ld} N$ bits, i. e. a message size of $O(\log N)$ during the concast phase is sufficient to achieve this. In the broadcast phase, each node v learns its height $h(v)$ and some starting slots to be described later. Given its height, it knows how many packets to relay that do not originate at nodes $u \in V_{k+2}$, i. e. $\sigma \left(D(v) \setminus V_{k+2} \right)$ and, if $h(v) \leq k + 2$, how many nodes to relay from each of the levels $h(v) + 1$ to $k + 2$. For the first, the pipelining phase, every node is additionally assigned a starting slot $T(v)$ as follows: The sink assigns itself the (imaginary) starting slot $T(r) = 0$, and using messages of at most $2\mathrm{ld} N$ bits, each node v assigns start slots $T(v) + 1 + \sum_{0 < j < i} \sigma \left(D(c_j) \setminus V_{k+2} \right)$ to each of its children c_i. For the second phase, nodes in V_{k+2} additionally receive the starting slots according to the scheme proposed in Proposition 2 restricted to nodes in V_{k+2} plus an offset of $(k+2) \cdot \sigma (V \setminus V_{k+2})$, which is known to the sink after the convergecast. In the pipelining phase, each node v but the sink now transmits in slots $T(v) + i(k+2)$ for $i = 0, \ldots, \sigma (D(v) \setminus V_{k+2})$. It receives a packet in the following slot the first $\sum_{c \in C(v)} \sigma (D(c) \setminus V_{k+2})$ times. This process is depicted in Fig. 2 for $k = 3$ and $\sigma \equiv 1$. Quite obviously, this part of the schedule has length $(k+2) \cdot \sigma (V \setminus V_{k+2}) + 1$. Nevertheless, for the analysis, we can assume a length of only $(k+2) \cdot \sigma (V \setminus V_{k+2})$ since the last transmission of this part can safely overlap with the first transmission of the second part. After completion, no node with height more than $k+2$ neighborhood has any packets left and each node v in V_{k+2} has exacly $\sigma (v)$ packets stored. Now, the rest of the packets is collected optimally as described in Proposition 2, i.e. in $\pi (V_{k+2})$ slots.

It is easy to see that the resulting schedule is energy optimal for the given tree. To show the approximation factor of $(k+2)/\lfloor (k+1)/2 \rfloor$, we observe that for $l := \lfloor (k+1)/2 \rfloor$, no two nodes in V_l can transmit at the same time. Thus, each packet originating at a node with height of more than l accounts for at least those l slots where it is the only one transmitted by a node with height of l or less in an optimal schedule. Similarly, every packet originating at a node $v \in V_l$ accounts for at least $h(v)$ slots. Hence, an optimal schedule has at least length

$$
\begin{aligned}
|S_{\text{OPT}}| &\geq l\sigma (V \setminus V_l) + \pi (V_l) \\
&= l\sigma (V \setminus V_{k+2}) + l\sigma (V_{k+2} \setminus V_l) + \pi (V_l)
\end{aligned}
\tag{4}
$$

The schedule produced by k-LS in turn uses

$$
\begin{aligned}
|S_{k\text{-LS}}| &= (k+2)\sigma (V \setminus V_{k+2}) + \pi (V_{k+2}) \\
&= (k+2)\sigma (V \setminus V_{k+2}) + \pi (V_{k+2} \setminus V_l) + \pi (V_l)
\end{aligned}
\tag{5}
$$

The claim now follows from the fact that $\pi (V_{k+2} \setminus V_l) \leq (k+2)\sigma (V_{k+2} \setminus V_l)$.

Applying very similar arguments as above, we can observe that first, k-LS produces a time optimal schedule if the first $(k+2)$ levels of T form a single path, and second, that if T is a hop-shortest path tree in the sink's l-hop neigborhood, the produced schedule is time approximative not only for an optimal schedule of T, but for an optimal schedule of any spanning tree. For the sake of brevity, we omit the proof.

Corollary 1. *The schedule produced by k-LS is time optimal for the given tree if V_{k+2} is a path and time approximative within a factor of $\frac{k+2}{\lfloor (k+1)/2 \rfloor}$ among all schedules of all spanning trees of G if $h(v) = d_G(v,r)$ for all v with $d_G(v,r) \leq l$, i.e. if T is a hop-shortest path tree in the l-hop neighborhood of r.*

All these results can very naturally be extended to a relaxed local interference model, the k_i/k_o-local interference model, in which a transmission (u, v) is successful if u is the only active sender in v's k_o-neighborhood, but might be successful if u is the only active sender in the k_i-neighborhood of v. Then, the approximation ratio from Theorem 2 becomes $\frac{k_o+2}{\lfloor (k_i+1)/2 \rfloor}$.

4   Making Schedules Robust

In this section, we will analyze an approach to make the above schemes robust to link failures using logarithmic sized additional memory per node. To allow for more sophisticated solutions than blindly repeating transmissions, which would not lead to a robust scheme anyway, we will assume some sort of ACK to acknowledge successful transmissions and a known lower bound on reception probability $\alpha(u)$ (including the feedback) for every link $(u, v) \in T$. We further assume reception probabilities to be independent for every transmission. We will first analyze a generic approach and then discuss individual issues of the proposed schemes.

For the generic approach, we will assume a non-robust schedule in which no node receives twice in a row without sending in between like all of the above. We further assume that we we have a mechanism to allocate $t(u) \geq \lceil 2/\alpha(u) \rceil$ successive slots for every scheduled transmission (u, v) in the non-robust schedule. We will discuss later how to do this for the individual protocols. The basic idea now is to run the non-robust schedule, with $t(u)$ slots reserved for every transmission (u, v). We will show that as long as every node only uses the $t(u)$ slots allocated for a single transmission in the non-robust schedule for at most one *first* transmission attempt (and arbitrarily many retransmissions), with high probability all packets are delivered if nodes are able to buffer a logarithmic number of packets and if the schedule budgets for a slightly more than $\sigma(u)$ packets for every node, namely $\sigma'(v) := \sigma(v) + \hat{\sigma}(v)$. More specifically, let every node v have a buffer of size of $(|C(v)| + 1)b$ for $b := 3 \log_4 N$ packets, and let $\hat{\sigma}(v) := (\frac{25}{32} \ln 4 + \frac{5}{2}(|C(v)| + 1))b$. We propose a robust transmission protocol (RTP) where a node u uses the $t(u)$ slots reserved for a transmission (u, v) in the non-robust schedule as follows: First, if all packets in the sender's buffer are marked as "pending", it pushes a "fresh" own packet to the buffer if there are any. Second, if there are unmarked packets in the buffer, it then marks one of them as "pending". Third, it tries to transmit as many pending packets as possible, removing successfully transmitted packets from the buffer. The receiver v simply adds new packets to its buffer if they have not been received before[2].

Proposition 4. *With high probability, RTP delivers all packets.*

Before we prove this claim, we prove the following lemma:

Lemma 1. *With high probability, no buffer of any node v ever contains more than b packets marked as pending or more than $|C(v)|b$ unmarked packets.*

Proof (Lemma 1). First, we observe that in a transmission phase, the number of nodes marked as pending at the sender can increase by at most 1, if all $t(u)$ transmission attempts fail and decreases by at least one if two or more transmission attempts are successful. Since the number of transmission attempts is

---

[2] Which can happen due to lost ACKs.

higher than $2/\alpha(u)$, the probability that the number of pending packets descreases, p^-, is more than four times the probability that this number increases, p^+, unless there were no pending packets at the begin of the transmission phase. Modeling the buffer utilization by marked packets as a finite markov chain, we get a probability of less than $(p^+/p^-)^b$ to be in a state where the buffer contains more than b marked packets by steady state analysis. The probability to reach such a state in any node in any transmission phase is thus less than

$$1 - \left(1 - (1/4)^b\right)^{N^2} = 1 - \left(1 - 1/N^3\right)^{N^2} < 1/N$$

On the other hand it is easy to see that also the number of unmarked packets in a node v's buffer can never exceed $|C(v)|b$: Looking at a node v and its children $C(v)$, we observe that the sum of packets marked as pending in the buffers of the children and the unmarked packets in the buffer of v can only increase during a transmission phase of a child and the next transmission phase of v (no node receives twice in a row) if v had no unmarked packet in its buffer prior to the transmission phase and the transmissions all failed. But since with high probability, the first number stays below b for all children, i. e. the sum of marked packets in the children's buffers does whp. not exceed $|C(v)|b$, the sum of marked packets at the children plus the number of unmarked packets at v cannot exceed $|C(v)|b$, which proves the claim.

Proof (Proposition 4). Before the additional $\hat{\sigma}(v)$ transmission phases start, every node had enough transmission phases to shift all its own packets to the buffer. From Lemma 1, we know that whp., the buffer sizes are sufficient. It remains to show that the additional $\hat{\sigma}(v)$ transmission phases are sufficient for every node to get rid of the packets left in the buffer, i. e. at most $(C(v) + 1)b$ packets. As argued above, in every transmission phase, with probability $p > 4/5$, at least one packet is transmitted. Thus, Hoeffding's inequality gives us the following upper bound on the probability that *for a single node*, less than $(|C(v)| + 1)b$ of the $\hat{\sigma}(v)$ calls are successful:

$$P_{\text{snf}} \leq \exp\left(-\frac{2(\hat{\sigma}(v)\,p - (|C(v)| + 1)b)^2}{\hat{\sigma}(v)}\right) < \exp\left(\frac{16(|C(v)| + 1)}{5}b - \frac{32}{25}\hat{\sigma}(v)\right)$$

With b and $\hat{\sigma}(v)$ as above, we get $P_{\text{snf}} < 1/N^3$, and a probability of less than $P_{\text{fail}} < 1 - (1 - 1/N^3)^{|V|} < 1/N$ that any node has any packets left when the schedule ends.

Note that while retransmissions of failed transmissions are inevitable, the proposed scheme does waste some energy for the following reasons: First, every node accounts for additional $O(\log N)$ packets, and second, nodes do have to transmit at least once during a transmission phase even if they do not have any packet marked as pending. Both effects increase the energy consumption only by small constant factors if every node has a payload of at least $\sigma(v) > \hat{\sigma}(v)$, i. e. $\sigma(v) \in \Omega(\log N)$. If, in this case, the number of time slots $t(v)$ can be set to exactly $\lceil 2/\alpha(v) \rceil$ for every v, the total number of time slots $2t(v)\sigma(D(v))$ reserved for v's transmission attempts is at most six times the expected number of necessary transmission attempts $\sigma(D(v))/\alpha(v)$.

Adapting this approach to the proposed schedulings, however, can incur additional costs. This adaption is easy if there is some reasonable lower bound on reception probability, i.e., if there is some small constant $c > 1$ for which $\max_{v\in V} \alpha(v) < c\min_{v\in V} \alpha(v)$. Then, every scheme can be made robust by choosing $t \equiv \lceil 2/\min_{v\in V} \alpha(v) \rceil$. Schemes in the total interference model can also be made robust by treating a link $(v, \text{parent}(v))$ with reception probability $\alpha(v)$ as if it was a path of length $t(v) = \lceil 2/\alpha(v) \rceil$. In this case, the height of the tree changes accordingly.

For the time-optimal scheduling scheme in the total interference model, robustness can cause higher costs: Since packets contain additional routing data, missing packets can sometimes only be compensated for by idle listening when waiting for the next packet. On the other hand, buffer sizes and packet count increase can be lowered for some of the proposed solutions: If a protocol guarantees that a node does not receive packets from children alternatingly, which does hold for all protocols but the level-order schemes, then the $|C(v)|$ can safely be replaced by 1 in the definition of b and $\hat{\sigma}(v)$.

5 Conclusion and Open Problems

We have analyzed the performance of different TDMA schemes for two interference models. Under the total interference model we have analyzed a scheme that optimizes the number of transmissions and the time to complete at the cost of increasing the packet size. We described another scheme which does not need to change the packet size but which is not time optimal. We have conjectured that there is no scheme which is time and energy optimal at the same time. The proof of this conjecture remains an open problem. It is also an open question if there are energy optimal schedules with better approximation guarantees (concerning time to complete) than $\Theta(\sqrt{n})$.

For the k-layer interference model we have proposed a scheme which is energy optimal and is a good approximation with respect to time till completion. Finally, we have shown how our schemes can be improved in order to integrate robustness mechanisms.

Our analysis has shown some lower bounds on the performance of TDMA schemes for data harvesting. It remains to evaluate the practical performance of our algorithm compared to other approaches under more realistic conditions.

Acknowledgements

The authors would like to thank Robert Görke and Reinhard Bauer for proofreading and fruitful discussions.

References

1. Xu, N., Rangwala, S., Chintalapudi, K.K., Ganesan, D., Broad, A., Govindan, R., Estrin, D.: A Wireless Sensor Network for Structural Monitoring. In: 2nd Int. Conf. on Embedded networked sensor systems (SenSys 2004), pp. 13–24. ACM Press, New York (2004)
2. Turau, V., Weyer, C.: Scheduling Transmission of Bulk Data in Sensor Networks Using a Dynamic TDMA Protocol. In: 8th Int. Conf. on Mobile Data Management (MDM 2007), pp. 321–325. IEEE Computer Society Press, Los Alamitos (2007)
3. Bermond, J.C., Galtier, J., Klasing, R., Morales, N., Perennes, S.: Hardness and Approximation of Gathering in Static Radio Networks. Parallel Processing Letters 16(2), 165–183 (2006)
4. Bonifaci, V., Korteweg, P., Marchetti-Spaccamela, A., Stougie, L.: An Approximation Algorithm for the Wireless Gathering Problem. In: Arge, L., Freivalds, R. (eds.) SWAT 2006. LNCS, vol. 4059, pp. 328–338. Springer, Heidelberg (2006)
5. Langendoen, K., Halkes, G.: Energy-efficient medium access control. In: Zurawski, R. (ed.) Embedded Systems Handbook. CRC Press, Boca Raton (2005)
6. Lu, G., Krishnamachari, B., Raghavendra, C.S.: An Adaptive Energy-Efficient and Low-Latency MAC for Data Gathering in Wireless Sensor Networks. In: 18th Int. Parallel and Distributed Processing Symp (IPDPS 2004), p. 224a. IEEE Computer Society, Los Alamitos (2004)
7. Hohlt, B., Doherty, L., Brewer, E.A.: Flexible power scheduling for sensor networks. In: 3rd Int. Symp. on Information Processing in Sensor Networks (IPSN 2004), pp. 205–214. IEEE Computer Society, Los Alamitos (2004)
8. Yao, Y., Alam, S.M.N., Gehrke, J., Servetto, S.D.: Network Scheduling for Data Archiving Applications in Sensor Networks. In: 3rd Worksh. on Data Management for Sensor Networks (DMSN 2006), pp. 19–25. ACM Press, New York (2006)
9. Turau, V., Weyer, C.: TDMA-Schemes for Tree-Routing in Data Intensive Wireless Sensor Networks. In: 1st Int. Work. on Protocols and Algorithms for Reliable and Data Intensive Sensor Networks (PARIS), pp. 1–6. IEEE Computer Society Press, Los Alamitos (2007)
10. Burri, N., von Rickenbach, P., Wattenhofer, M.: Dozer: Ultra-Low Power Data Gathering in Sensor Networks. In: 6th Int. Symp. on Information Processing in Sensor Networks (IPSN 2007), pp. 450–459. ACM Press, New York (2007)

Link Scheduling in Local Interference Models

Bastian Katz⋆, Markus Völker, and Dorothea Wagner

Faculty of Informatics, Universität Karlsruhe (TH)
{katz,mvoelker,wagner}@ira.uka.de

Abstract. Choosing an appropriate interference model is crucial for link
scheduling problems in sensor networks. While graph-based interference
models allow for distributed and purely local coloring approaches which
lead to many interesting results, a more realistic and widely agreed on
model such as the signal-to-noise-plus-interference ratio (SINR) inherently makes scheduling radio transmission a non-local task, and thus
impractical for the development of distributed and scalable scheduling
protocols in sensor networks. In this work, we focus on interference models that are *local* in the sense that admissibility of transmissions only
depends on local concurrent transmissions, and *correct* with respect to
the geometric SINR model.

In our analysis, we show lower bounds on the limitations that these
restrictions impose an any such model as well as approximation results
for greedy scheduling algorithms in a class of these models.

1 Introduction

Agreeing on good schedules in wireless networks is not only a question of good,
i. e., local and distributed scheduling algorithms. The correctness of any scheduling algorithm's output relies on the underlying interference model. Choosing an
interference model is thus crucial for any kind of scheduling protocol in sensor
networks. Both, interference models and scheduling problems have been studied
thoroughly in the "tradition" of sensor networks. Complex interference models incorporating sophisticated signal fading models and antenna characteristics,
developed over the years, proved helpful in the simulation and design of sensor
networks. In the algorithmic community, however, the need of clear, preferably
combinatoric and geometric interference models, led to a focus on graph-based
interference models. These graphs all have in common that they are local in the
sense that mutual exclusion between transmissions only "connects" nodes that
are close to each other. The simple combinatorial character of these models naturally translates scheduling problems to coloring problems in graphs. Moreover,
the geometric properties of these graphs allow for tailored coloring protocols.

Despite their simplicity, the downside of these models clearly is that they could
neither be proven to be correct nor good in real sensor networks. They cannot

---

⋆ Partially supported by the German Research Foundation (DFG) within the Research
Training Group GRK 1194 "Self-organizing Sensor-Actuator Networks".

S. Fekete (Ed.): ALGOSENSORS 2008, LNCS 5389, pp. 57–71, 2008.

model interference from far away nodes summing up and jamming communication, nor can they model that if in reality any pair out of three transmissions can successfully be performed simultaneously, this does not necessarily mean that all three transmissions can be performed simultaneously.

Algorithmic research considering a class of models that renders signal propagation much more realistically did to the best of our knowledge not yet lead to local algorithms. In SINR models, successful or sufficiently probable reception is assumed if at a receiver, the respective sender's signal strength outperforms the sum of all interfering signals plus the background noise by a hardware dependent constant. The geometric SINR models closely cover the main features of sophisticated fading models such as the two-ray-ground model without losing too much of the simplicity needed for algorithmic results.

In this paper, we introduce the concept of locality and correctness of interference models. We prove fundamental limitations of all models that are local in a very straightforward sense and correct with respect to the geometric SINR model. We show under which conditions well known concepts such as graph coloring *can* be used to approximate scheduling problems and a generalization that improves the quality of easy-to-implement scheduling algorithms. We believe that the introduced models open a door to more realistic, yet viable solutions not only for scheduling, but for many protocols that rely on local, dependable communication.

2   Related Work

Interference of concurrent communication, being the most outstanding attribute of wireless networks, has been subject of countless publications. Since in reality, interference is composed of many hard-to-capture phenomena such as multipath fading, algorithmic research developed numerous simplifications. Most of the algorithmic models model interference as a binary relation on transmissions, among them the unit disk graph (UDG) with distance or hop interference or the protocol model. We refer the reader to [1] for a survey. In SINR models, successful reception depends on the ratio between the received signal strength on the one side and the interference from concurrent transmissions plus the background noise on the other side [2,3]. They differ in whether they assume signal strength decay to be a function of the distance (geometric SINR) or allow an arbitrary gain matrix. In the geometric SINR model, Gupta and Kumar analyzed the capacity of ad-hoc networks and proved an upper bound on the throughput of $\Theta(1/\sqrt{n})$ for networks of n nodes. Until now, the effects of the SINR models to algorithm design raise interesting questions [4].

Scheduling of link transmissions has been addressed in many interference models and, in most cases, proven to be NP-hard. Among others are proofs for scheduling in graph-based models [5,6], in the abstract SINR model in [7], and, recently, in the geometric SINR model for fixed power assignment by Goussevskaia et al. [8]. The joint problem of scheduling and power assignment is still open in the geometric SINR model [9]. A variety of graph-based scheduling algorithms has been proposed and analyzed [10, 11, 12]. It is however argued in various works that graph-based scheduling is inferior to scheduling designed for the SINR model [13, 14]. Among the early publications addressing scheduling in geometric SINR models, Moscibroda and Wattenhofer show that uniform or linear power assignments in worst-case scenarios need exponentially longer schedules for a strongly connected set of links [15] than more sophisticated assigments. Moscibroda et al. also propose a scheduling algorithm for arbitrary power assignment in [16] that outperforms previous heuristics by an exponential factor. In [8], Goussevskaia et al. propose an approximation algorithm for link scheduling and the problem of finding a maximum number of links that can transmit concurrently in the geometric SINR model under the fixed power assumption. The latter three works introduced many of the techniques applied in the following under the practically more relevant assumptions that nodes do not feature arbitrarily high transmission powers and cannot rely on a global instance to compute a schedule, but are restricted to a local view. Locality has, to our knowledge, only been looked at in a combinatorial sense [17, 18, 19].

3   Definitions and Models

A *deterministic* interference model $\mathcal{M}$ is a property telling for a fixed set of nodes $\mathcal{V}$ whether a set of transmissions T between nodes in $\mathcal{V}$ can be carried out simultaneously for given transmission powers. More formally, let $\mathcal{T} := \left\{ (u, v, p) \in \mathcal{V}^2 \times \mathbb{R}_+ \mid u \neq v \right\}$ be the set of all possible transmissions and transmission powers. Then, a model $\mathcal{M} \subseteq \mathcal{P}(\mathcal{T})$ contains all sets of transmissions which are valid. We further assume that less concurrent transmissions cannot cause a transmission to fail, i.e., that for all $T' \subseteq T \subseteq \mathcal{T}$,

$$T \in \mathcal{M} \Rightarrow T' \in \mathcal{M} \ , \tag{1}$$

which holds for all models which are currently used and most likely for all models which are meaningful. One should note that the restriction to deterministic models alone already is a giant step away from reality and the probabilistic models typically employed by communication theorists. But still, even deterministic models are not understood well. Such models can rely on various kinds of additional input and assumptions of radio propagation, antenna characteristics and so on. In higher layer protocol design, however, there is a need to "model away" the complexity of most of these unrulable phenomena. An aspect that has not received much attention yet is how different approaches to model interference relate to each other, or, in other words: If I choose a simpler model, are my algorithms or schedules still correct with respect to a more realistic one or are they just suboptimal? Do optimal solutions in a simple model approximate optimal solutions in a more complex model? In the following, we will call an interference model $\mathcal{M}$ *conservative* with respect to another model $\mathcal{M}'$ if $\mathcal{M} \subset \mathcal{M}'$.

Most analytical research on scheduling problems has been done in some kind of *graph-based* interference model accordig to the following definition from [4].

Definition 1 (Graph-Based Model). *A graph-based model $\mathcal{M}$ can be defined by two directed graphs, one connectivity graph $D_C = (\mathcal{V}, A_C)$ restricting possible transmissions and one interference graph $D_I = (A_C, A_I)$ connecting conflicting transmissions, such that $T \in \mathcal{M}$ if and only if $T \subset A_C$ and $T^2 \cap A_I = \emptyset$.*

Usually, a simpler model consisting of two graphs $G_C = (\mathcal{V}, E_C)$ and $G_I = (\mathcal{V}, E_I)$ is used, in which a set of transmissions is valid, if for every sender, the intended receiver is a neighbor in G_C and no receiver of a distinct transmission is connected in G_I. Sometimes, the connectivity graph and interference graph are defined implicitly, i. e., as the result of a geometric setting.

Graph-based models all have in common that they claim that a set of transmissions whose transmissions can *pairwise* be carried out at the same time, collectively may be scheduled into one single time slot. This is unrealistic in general, and the models fail to formulate the assumptions under which they guarantee not to produce schedules that do not comply with more realistic models. On the other hand, in the single-power case, graph-based models reduce scheduling problems to well-known coloring problems.

As opposed to the oversimplification of graph-based interference models, the models capturing the findings of signal propagation and reception best are the *signal-to-noise-plus-interference* (SINR) models. Their main paradigm is that a transmission is (almost) always successful, if the sender's signal strength *at the receiver* is significantly stronger than the sum of all interfering signals, including other sender's signals and (individual) background noise. Thus, in its most general form, an SINR model is defined by a *gain* matrix (G_{uv}) denoting the signal fading between nodes u and v, on the background noise η_v at each of the nodes and the (individual) ratio β_v a node v needs for proper reception. Here, a set of transmissions is valid, i. e. $T \subset \mathcal{M}$, if and only if for all $t = (s, r, p_s) \in T$

$$\frac{p_s G_{sr}}{\eta_r + \sum_{(u,v,p_u) \in T \setminus \{t\}} p_u G_{ur}} \geq \beta_r \ . \tag{2}$$

Definition 2 (Geometric Model). *In a geometric model, $\mathcal{M}$ is defined for $\mathcal{V} = \mathbb{R}^2$ such that $\mathcal{M}$ is invariant under all isometries.*

Generally speaking, geometric interference models are incapable of modeling individual characteristics of nodes, but are restricted to those of geometric settings. This does not mean that a geometric model has to be parameter-free, but for geometric SINR models, this definition implies a much simpler structure:

Theorem 1. *Every geometric SINR model can also be defined equivalently such that all η_v and all β_v are independent of the respective position v and all G_{uv} can be expressed as $G_{uv} := f(d(u, v))$ for a $f : \mathbb{R} \to \mathbb{R}$.*

Proof. Let $\mathcal{M}$ be a geometric SINR model defined by (G_{ji}), (η_i) and (β_i). We get an equivalent model for (G'_{ji}), (η'_i), and (β_i), in which all $\eta'_i = \eta'$ are the same by setting $G'_{ji} = G_{ji} \frac{\eta'}{\eta_i}$. Now, for any d, take two pairs u_1, v_1 and u_2, v_2. We know that $\{(u_i, v_i, p)\} \in \mathcal{M}$ if and only if $p \geq \eta' \beta'_{v_i} / G_{u_i v_i}$. Since there is an isometry mapping u_1 to u_2 and v_1 to v_2, and since $\mathcal{M}$ is geometric, transmissions (u_1, v_1, p) are valid, for exactly the same values of p as (u_2, v_2, p). Thus, $p \geq \eta' \beta'_{v_1} / G'_{u_1 v_1}$ if and only if $p \geq \eta' \beta'_{v_2} / G'_{u_2 v_2}$ and thus $\beta'_{v_1} / G'_{u_1 v_1} = beta'_{v_2} / G'_{u_2 v_2}$. I. e., all pairs of nodes with distance d have the same ratio of G'_{ji} and β_i and by fixing some β' and setting $G''_{ji} = G'_{ji} \beta' / \beta_i$, we get a representation of the claimed form.

The class of *geometric SINR models (SINR$_G$)* is a quite straightforward application of the above definition. Individual characteristics such as the background noise and the necessary SINR ratio are replaced by common constants η and β and the gain G_{uv} is replaced by a function of the distance, usually $K d_{uv}^{-\alpha}$ for a so-called *path-loss exponent* α and some constant K. Currently, the SINR$_G$ models widely agreed are the best models to reason about in the algorithmics of sensor networks. Thus, we will focus on local models that are conservative with respect to this class of models.

A *scheduling problem* in a wireless network is a set $\mathcal{Q}$ of communication requests, each request (s, r) consisting of a sender s and a receiver r, both from some set $\mathcal{V}$ of nodes. A schedule then is a sequence $T_1, T_2, \ldots, T_k$ of sets of transmissions of the form (s, r, p) for some $(s, r) \in \mathcal{Q}$ and some power assignment $p \in \mathbb{R}_+$, such that for every $(s, r) \in \mathcal{Q}$, there is a transmission (s, r, p) in one of the T_i, and every T_i is valid with respect to an interference model. We refer to the problem of finding a schedule of minimum length as SCHEDULE, and to the problem of finding a maximum number of transmissions that can be scheduled to a single slot as ONESHOTSCHEDULE as in [8]. We will also denote the maximum link lenght occurring in a schedule request $\mathcal{Q}$ by $\ell(\mathcal{Q}) := \max_{(s,r) \in \mathcal{Q}} d_{sr}$. If the scheduling problem is combined with the problem of assigning transmission powers, usually powers must be chosen from some power range $\mathbf{p} = [p_{\min}, p_{\max}]$. In the following, we will focus on the problem of finding schedules for a fixed power p and thus also write (s, r) to denote a transmission (s, r, p).

4   Local Interference Models

The concept of locality has been introduced for distributed systems and adopted in the context of sensor networks. Usually, a distributed algorithm is said to be k-local, if the outcome for every node only depends on nodes which are inside a k-hop-neighborhood. Unfortunately, this concept is too restrictive to allow for any local scheduling algorithms in a geometric SINR model with nodes that do not feature arbitrarily high transmission powers, but are limited to some maximum power. Even if we define a node's neighborhood as the set of nodes the node can communicate with when no other communication takes place at the same time, in an SINR$_G$ model $\mathcal{M}_G$, it might be impossible to arrange a schedule at all. If we denote the maximum possible link length of an interference model $\mathcal{M}$ by $\ell(\mathcal{M}) := \limsup_{\{(u,v,p)\} \in \mathcal{M}} d_{uv}$, we get $\ell(\mathcal{M}_G) = \sqrt[\alpha]{Kp/(\beta\eta)}$, since for nodes with higher distance even in the absence of concurrent transmissions sending at maximum power does not result in a received signal strength of $\beta\eta$, which is necessary due to the background noise alone.

In Fig. 1, such a situation is depicted: Out of the two
sender/receiver pairs in the transmission request, only the
pairs themselves have a distance less than $\ell(\mathcal{M}_G)$, and
thus, there is no communication possible between the different pairs, which, however, have to agree not to transmit
at the same time since both of them cannot compensate
for the interference caused by the other. We will thus in
the next section look at a weaker, geometric definition of
locality and its consequences for scheduling problems.

Fig. 1. Links with dis-
tance $\ell(\mathcal{M}_G) + \epsilon$ may
need communication

Definition 3 (Local Model). *A ρ-local model is a geometric model $\mathcal{M}$ with
the additional constraint that $T \in \mathcal{M}$ if for every $t = (s, r, p) \in T$*

$$T(s, \rho) := \{(s', r', p') \in T \mid d(s, s') \leq \rho\} \in \mathcal{M} \ .$$

In other words, an interference model is *local*, if for a set of transmissions T, it
is sufficient that for every sender in T the transmissions in its ρ-neighborhood
comply with the model to make T valid. Models of this kind not only allow to tell
that a set of transmissions will be successful by only locally looking at the transmissions, but they are also essential for the design of local algorithms. They can
be seen as a rule for every node that can only observe nearby nodes, either during a setup phase or, more importantly maintaining a dynamic link transmission
schedule. The geometric graph-based models mentioned above quite naturally
have this property, but SINR_G models do not, which proved to be one of the
main obstacles when tackling scheduling problems in these models. This holds
for existing centralized approximation algorithms which try to break the inter-
woven dependencies into independent subproblems as in [8], and it inherently
does so in distributed settings – how could nodes come up with a provably valid
schedule with local communication, when the validity of a schedule cannot be
judged locally? Local interference models on the other hand seem to be incorrect by design: They are blind for interference that arises from nodes that are
far away, and thus cannot factor what these nodes are doing. From this time
on, let $\mathcal{M}_G = (K, \eta, \beta, \alpha)$ be a standard SINR_G model. We start with an observation which illustrates the first limitations local reasoning about interference
implicates. It is a generalization of the considerations above.

Observation 1. *Let $\mathcal{M}_L$ be a $\mathcal{M}_G$-conservative, ρ-local interference model. The
following two inequalities hold:*

$$\ell(\mathcal{M}_L) < \frac{\rho}{1 + \sqrt[\alpha]{\beta}} \quad and \quad \ell(\mathcal{M}_L) < \ell(\mathcal{M}_G) \cdot \left(1 + \frac{3Kp}{\eta\rho^\alpha}\right)^{-1/\alpha} \tag{3}$$

Proof. Let $T \in \mathcal{M}_L$ be any set of transmissions that is accepted by the local
model and $t = (s, r)$ any transmission in T. Let $\ell_t := d_{sr}$. By the subset acceptance property (1) and the fact that $\mathcal{M}_L$ is geometric (and thus invariant under
isometries of the plane), $\mathcal{M}_L$ would accept any set of transmissions T in which
all senders have pairwise distances of more than ρ and all transmissions from T
have length ℓ_t.

Now, consider a set of transmissions T, as depicted in Fig. 2, where senders are placed on a triangular grid with edge length $\bar{\rho} := \rho + \epsilon$, i.e., which complies with the considerations above. First, if we assume that $\ell_t \geq \rho/(1 + \sqrt[\alpha]{\beta})$, the interference of the sender s_2 alone would interfere with the reception of the transmission t to the limit,

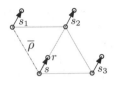

Fig. 2. Lower bounding interference in ρ-local models

$$\lim_{\epsilon \to 0} \frac{Kp\bar{\rho}^{-\alpha}(1 + \sqrt[\alpha]{\beta})^\alpha}{Kp\bar{\rho}^{-\alpha}(1 - 1/(1 + \sqrt[\alpha]{\beta}))^{-\alpha}} = \beta \ ,$$

and together with the additional interference caused by other senders, reception would become impossible, contradicting with the choice of t.

Second, as we now know that $\ell_t < \rho/2$, we get that the interference of the senders s_1, s_2 and s_3 at the receiver is at least $3Kp\rho^{-\alpha}$. Thus, since $T \in \mathcal{M}_G$,

$$\frac{Kp\ell_t^{-\alpha}}{3Kp\rho^{-\alpha} + \eta} \geq \beta \Leftrightarrow \ell_t \leq \left(\frac{\beta\eta}{Kp} + \frac{3\beta}{\rho^\alpha}\right)^{-1/\alpha} = \ell(\mathcal{M}_G) \cdot \left(1 + \frac{3Kp}{\eta\rho^\alpha}\right)^{-1/\alpha} \ ,$$

which concludes the proof.

Note that this bound is by no means tight, but it shows how severe the restrictions are that one can only overcome by globally solving schedule problems: To allow for longer links, especially of lengths close to $\ell(\mathcal{M}_G)$, the radius ρ has to be chosen accordingly. We can derive better bounds by calculating interferences more accurately than above by summing up interference for more senders on the same triangular grid. Fig. 3 shows an

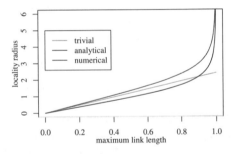

Fig. 3. Lower bounds on ρ ($\alpha = \beta = 4.0$)

exemplary tradeoff between the maximum link length needed, $\ell(\mathcal{M}_L)$ and the resulting analytical and numerical lower bounds on the radius ρ for $\alpha = 4$, $\beta = 4$ ($\approx$ 6dB) and η, p, K normalized to $\ell(\mathcal{M}_G) = 1$. It shows that in the case that we do not assume that nodes can communicate with nodes outside their transmission radius, e.g., by the assumption that the node density is sufficiently high, no link length longer than 40% of the maximum link length can safely be scheduled in realistic scenarios.

The second observation we can make about local interference models regards the case that nodes cannot send with arbitrarily low power:

Observation 2. *Let $\mathcal{M}_L$ be a $\mathcal{M}_G$-conservative ρ-local interference model for a $\rho < \infty$. Even for requests with $\ell(\mathcal{Q}) \leq \ell(\mathcal{M}_L))$, optimal solutions to* SCHEDULE *and* ONESHOTSCHEDULE *in $\mathcal{M}_L$ can be arbitrarily worse than in $\mathcal{M}_G$.*

For the sake of brevity, we will only give a sketch of the proof here. We look at a request of a ring of n transmissions as depicted in Fig. 4 with sufficiently small transmission lengths $\ell(g)$ plus one transmission $t^\star$ of length $\ell(n)$ in the middle. It is easy to see that in $\mathcal{M}_G$, it is admissible to schedule all transmissions but $t^\star$ to the same slot (and $t^\star$ to a second). In $\mathcal{M}_G$, assigning a slot to $t^\star$ and to the rest of transmissions is independent. Thus, at no time more than a constant number of the n outer transmissions can be carried out, allowing for concurrent transmission of $t^\star$.

Fig. 4. Ring of transmissions

5 $\Omega(1)$-Sender Model

In every meaningful local model, acceptance of a (local) set of transmissions must follow this consideration: Given the rules for local acceptance of a set of transmissions – is it guaranteed that if all nodes obey these local rules, no node possibly has to accept more interference from outside the ρ-neighborhood than allowed, given the amount of interference arising from local transmissions. A quite straightforward implementation of this concept is the following: For some function $\mu : \mathbb{R}_+ \to \mathbb{R}_+$, which serves as an upper bound for interference from far away nodes, a set of transmissions T is licit if for every transmission (s,r) a local *signal-to-noise-plus-interference* condition holds:

$$\frac{Kpd_{sr}^{-\alpha}}{\sum_{(\hat{s},\hat{r})\in T(s,\rho)} Kpd_{\hat{s},r}^{-\alpha} + \eta + \mu(d_{sr})} \geq \beta \ , \tag{4}$$

and if it is guaranteed that a transmission (s,r) cannot receive more interference than $\mu(d_{sr})$ from senders further away than ρ from s. One way to guarantee the latter is to prohibit that close senders are transmitting concurrently and thus, to limit the density of active senders:

Definition 4 ($\Omega(1)$-sender model). *In the $\Omega(1)$-sender model $\mathcal{M} = (\rho, c, \mu)$, a set of transmissions T is valid if and only if for every $(s,r) \in T$ equation (4) holds, and any two senders in T have distance at least c.*

Such a model clearly is ρ-local if $c \leq \rho$, but quite obviously not $\mathcal{M}_G$-conservative for an arbitrary μ. However, for certain values of ρ, c, and μ, the resulting model (ρ, c, μ) is $\mathcal{M}_G$-conservative and local:

Lemma 1. *Let $\mathcal{M}_L = (\rho, c, \mu)$ be an $\Omega(1)$-sender model. $\mathcal{M}_L$ is conservative with respect to $\mathcal{M}_G$ if* [1]

$$\mu(\ell) \geq \frac{\sqrt{12}Kp\pi\zeta(\rho^2/c^2 + 2\rho/c)}{(\rho - \ell)^\alpha} =: \mu_1(\ell) \tag{5}$$

---

[1] For the Riemannian ζ-function and $\zeta := \zeta(\alpha - 1)$, a constant $1 < \zeta < 2$ for $\alpha \geq 3$.

Proof. Let (s, r) be some sender/receiver pair with $d_{sr} = \ell$. We divide the plane into annuli A_k with center s and radii $k\rho$ and $(k+1)\rho$ for $k \in \mathbb{N}$. The maximum number of senders lying within the kth annulus is the maximum number of disks of radius c within an annulus with radii $k\rho - c$ and $(k+1)\rho + c$. Since senders in $\Omega(1)$-sender models form a Minkowski arrangement, which cannot exceed a density of $\frac{2\pi}{\sqrt{3}} \approx 3.638$ [20], we get that the number of senders in A_k is at most

$$N_k := \left\lfloor \frac{2\pi}{\sqrt{3}} \cdot \frac{\pi\left(((k+1)\rho + c)^2 - (k\rho - c)^2\right)}{\pi c^2} \right\rfloor \leq k \underbrace{\sqrt{12}\pi \left(\rho^2/c^2 + 2\rho/c\right)}_{:=N^*},$$

The interference received from any of the senders in A_k can be bounded by

$$I_k := Kp(k\rho - \ell)^{-\alpha} \leq k^{-\alpha} \underbrace{Kp\,(\rho - \ell)^{-\alpha}}_{=:I^*},$$

and the total interference received from any sender can then be bounded by $\sum_{k=1}^{\infty} N_k I_k \leq N^* I^* \sum_{k=1}^{\infty} k^{-\alpha+1} = N^* I^* \zeta$.

Let (ρ, c) denote a shortcut for the $\mathcal{M}_G$-conservative model (ρ, c, μ_1). Obviously, the bound μ_1 can very straightforward be replaced by a better numerical bound. Fig. 5(a) shows how these bounds compare to each other and to the lower bound from the last section. Note that all bounds are correct, and, given the SINR parameters, easy to calculate. We will use the closed-form result for further analysis and the improved bounds for simulation.

With the approximation above, we still have the choice of first the maximum possible link length and second, the balance of the locality factor ρ and the exclusion radius c.

Corollary 1. *Let $\ell = u \cdot \ell(\mathcal{M}_G)$ for some $0 < u < 1$ be a link length and $a \geq 1$, the $\mathcal{M}_G$-conservative $\Omega(1)$-sender model $\mathcal{M}_L = (\rho, \rho/a)$ with*

$$\rho = \left(1 + \sqrt[\alpha]{\frac{\sqrt{12}(a^2 + 2a)\pi\beta\zeta}{1 - u^\alpha}}\right) \cdot \ell$$

has $\ell(\mathcal{M}_L) = \ell$. It is graph-based for $a = 1$, yielding (ρ, ρ).

Proof. According to Lemma 1, $\mathcal{M}_L$ is $\mathcal{M}_G$-conservative. It remains to show that $\ell(\mathcal{M}_L) = \ell$. First, by $\ell = u \sqrt[\alpha]{Kp/\beta\eta}$, we observe that

$$\rho - \ell = \sqrt[\alpha]{\frac{\sqrt{12}(a^2 + 2a)\pi\beta\zeta}{1 - u^\alpha}} \cdot \ell = \sqrt[\alpha]{\frac{\sqrt{12}(a^2 + 2a)\pi Kp\zeta}{\eta(u^{-\alpha} - 1)}} \tag{6}$$

and therefore by the definition of u

$$\ell(\mathcal{M}_L) = \sqrt[\alpha]{\frac{Kp}{\beta\eta + \sqrt{12}(a^2 + 2a)\pi Kp\beta\zeta\,(\rho - \ell)^{-\alpha}}} = \sqrt[\alpha]{\frac{Kp}{\beta\eta u^{-\alpha}}} = \ell$$

Obviously, it is graph-based for $a = 1$.

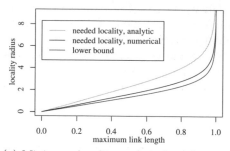
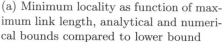
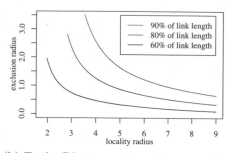

(a) Minimum locality as function of max-imum link length, analytical and numeri-cal bounds compared to lower bound

(b) Tradeoff between the minimum pair-wise sender distance and the locality ra-dius for fixed $\ell(\mathcal{M}_L)/\ell(\mathcal{M}_G)$

Fig. 5. Tradeoffs for $\Omega(1)$-sender model (ρ, ρ)

This corollary in a way justifies the work that has been done on scheduling problems in graph-based models as it provides a very simple graph-based model that is provably correct with respect to the geometric SINR models (and, at the same time shows the price for reducing an SINR_G model to a graph-based model). Fig. 5(b) shows this tradeoff between the maximum schedulable link length and the respective ρ for different ratios. As argued in Section 4, no local model can allow for "good" solutions in the single-power setting in the sense that the scheduling problems can be approximated within a constant factor in any such model if links can be arbitrarily short. We will show that $\mathcal{M}_G$-conservative $\Omega(1)$-sender models (ρ, ρ) are only by constant and comparably small factors worse than any local model in two dimensions – first the locality needed to allow for a given maximum link length and second the quality of optimal solutions to the scheduling problems.

Lemma 2. *Let* $\mathcal{M}_L = (\rho, \rho)$ *be the* $\mathcal{M}_G$-*conservative* $\Omega(1)$-*sender model ac-cording to Corollary 1 for some* ℓ. *Then for any* ρ'-*local model* $\mathcal{M}'_L$ *which is* $\mathcal{M}_G$-*conservative,*

$$\frac{\rho}{\rho'} \leq \frac{1}{\sqrt[\alpha]{3\beta}} + 2\sqrt{3}\pi\zeta$$

Proof. From (3) and with $u := \ell/\ell(\mathcal{M}_G)$, we get that

$$\rho' = \sqrt[\alpha]{3\beta} \left(\frac{1}{\ell(\mathcal{M}_L)^\alpha} - \frac{1}{\ell(\mathcal{M}_G)^\alpha} \right)^{-1/\alpha} = \ell \sqrt[\alpha]{3\beta} \left(1 - u^\alpha \right)^{-1/\alpha}$$

and, from Corollary 1 and $u < 1$,

$$\rho = \left(1 + \sqrt[\alpha]{\frac{6\sqrt{3}\pi\beta\zeta}{1 - u^\alpha}} \right) \cdot \ell < (1 - u^\alpha)^{-1/\alpha} \cdot \left(1 + \sqrt[\alpha]{6\sqrt{3}\pi\beta\zeta} \right) \cdot \ell ,$$

which directly implicates the claimed approximation.

This bound depends only on α and β, and is thus constant for a fixed SINR_G model. For a given set of SINR_G parameters, this approximation ratio can be improved using the non-closed-form lower bounds for local models and upper bounds for the c-distant sender model. E. g., for the exemplary values used throughout this paper, the best bounds guarantee a ratio of less than $5/4$ for arbitrary ℓ.

Lemma 3. *Let $\mathcal{M}_L = (\rho, \rho)$ be the $\mathcal{M}_G$-conservative $\Omega(1)$-sender model according to Lemma 1 and Corollary 1. Let $\mathcal{Q}$ be a schedule request with $\ell(\mathcal{Q}) < \ell(\mathcal{M}_L)$. Optimal solutions to* SCHEDULING *and* ONE-SHOT-SCHEDULE *in $\mathcal{M}_L$ are only by a constant factor worse than optimal solutions in any other ρ-local $\mathcal{M}_G$-conservative model.*

Proof. First we show that any ρ-local model $\mathcal{M}'_L$ with $\ell(\mathcal{M}'_L) \geq \ell$ cannot accept any set of transmissions T such that for any transmission (s, r, p), $T(s, \rho)$ contains more than $h := 4^\alpha 6\sqrt{3}\zeta$ transmissions. To this extent, let T be a set of h transmissions such that $T(s, \rho) = T$ for some $(s, r) \in T$. We add a transmission (s', r') to T with $d_{s,s'} = 2\rho$, pointing towards s. Note that if T is valid in $\mathcal{M}'_L$, then $T' = T \cup \{(s', r')\}$ must be valid, too. But if all transmissions in T' are carried out simultaneously, the SINR-level at r' is below

$$\frac{Kp\ell^{-\alpha}}{hKp(2\rho)^{-\alpha} + \eta} = \frac{\ell^{-\alpha}}{h4^{-\alpha}\frac{\ell^\alpha(1-u^\alpha)}{6\sqrt{3}\pi\beta\zeta} + \frac{\eta}{Kp}} = \frac{\ell^{-\alpha}}{\frac{\ell^{-\alpha}(1-u^\alpha)}{\beta} + \frac{\eta}{Kp}} = \beta \ .$$

Now take any schedule request $\mathcal{Q}$. Let A be the square with side $\rho/\sqrt{2}$ that contains the most senders in $\mathcal{Q}$. Let m denote this number. Since in every ρ-local model at most h of the senders in A can transmit concurrently, leading to a schedule length of at least $\lceil m/h \rceil$. In $\mathcal{M}_L$, in turn, we can construct a schedule of length $4m$ by the same construction as in [8]: We extend the square to a grid of grid-length $\rho/\sqrt{2}$, 4-color the grid cells, and cyclically choose a color and pick an unscheduled sender from each cell with that color. This guarantees a $4h$-approximative schedule compared to an optimal solution in any ρ-local model. Similar arguments lead to a $4h$-approximation of ONE-SHOT-SCHEDULE: Take an optimal solution T in any ρ-local model and picture a 4-colored grid with grid-length $\rho/\sqrt{2}$. Focus only on grid-squares that contain a sender of T. Each of the squares contains at most h transmissions in T. Now pick the color of the most non-empty squares and pick one transmission of each square. This set of transmissions contains at least $\lceil |T|/4h \rceil$ transmissions that can all be carried out concurrently in $\mathcal{M}_L$. We get a slightly worse approximation for greedy scheduling, where we will only look at the SCHEDULING problem: In the very same grid as above, if one cell does not contain an active sender in a slot, then for two possible reasons: First, since all senders have been scheduled to earlier slots, and second, because of some active sender in one of the adjacent cells. Thus, greedy scheduling uses at most $9m$ slots, which is $9h$-approximative.

6   Implementation and Simulation Results

We implemented a very basic scheduling algorithm for $\Omega(1)$-sender models which greedily assigns slots to senders in a random order: Each sender is assigned to the first allowed slot according to the respective model. This is not only a very simple centralized approach, but also a reasonable distributed scheduling algorithm. Given that the node density is sufficient for nodes to have their ρ-neighborhood some constant number of hops away, nodes can draw random numbers and decide on their slot after all neighbors with lower numbers did so only by local communication. This approach is also suited to schedule online requests. We compare the results to three different global scheduling algorithms. First, we select nodes in random order and add them to the first slot allowed by the plain SINR_G model. Second, we compare to the algorithm given in [8]. Please note that this algorithm is not designed to produce good schedulings, but only as a proof of approximability. It is thus not surprising that it returns comparably poor results. Third, since solving the SCHEDULE problem optimally is hard, and solving the corresponding mixed-integer linear problem only works for a very small number of transmissions, we compare to a heuristic, which produced near-optimal results for small instances of random transmission requests. We fill the slots one after another, at any time adding the transmission which causes the least drop of the minimum signal-to-noise-plus-interference ratio for all transmissions earlier added to that slot. We ran all of the above algorithms on schedule requests with at most 80% of the maximum link length in the SINR_G model. Instances were random sets of 20000 transmissions and random unit disk graphs with 5000 nodes on a 50x50 square unit area, i.e. some 10000 edges leading to some 20000 transmissions to schedule links symmetrically. For the $\Omega(1)$-sender models, we compare three configurations. First, the graph-based (ρ, ρ) model with minimum ρ, second a $(\rho, \ell(M_G))$ model with minimum ρ and third, a $(\rho, \ell(M_G))$ model for a ρ higher than necessary. Additionally, we compare to a variant of the $\Omega(1)$-sender models, where the locality radius ρ is centered at the receiver, where the interference occurs. We call this receiver-centered locality. This class of models never is

Table 1. Comparison of scheduling in different interference models

Algorithm	locality $[\ell(M_G)]$	excl. rad. $[\ell(M_G)]$	random links length	random links max. util.	random links avg. util.
Greedy scheduling in M_G	∞	0	71.91	482.45	278.43
Intelligent scheduling in M_G	∞	0	34.15	2726.00	586.11
Goussevskaia et al. [8]	∞	0	605.03	206.01	33.06
Greedy scheduling in ...					
... (ρ, ρ), min. ρ (graph-based)	2.82	2.82	117.38	224.31	170.45
... $(\rho, \ell(M_G))$, min. ρ	4.59	1.00	58.63	626.59	341.31
... $(\rho, \ell(M_G))$, incr. ρ	6.00	1.00	47.89	629.41	417.86
... rc.-local (ρ, r), min. ρ	2.55 (rc)	1.75	60.88	487.68	328.79
... rc.-local $(\rho, \ell(M_G))$, min. ρ	3.82 (rc)	1.00	54.84	675.81	364.91
... rc.-local $(\rho, \ell(M_G))$, incr. ρ	6.00 (rc)	1.00	46.48	646.45	430.50

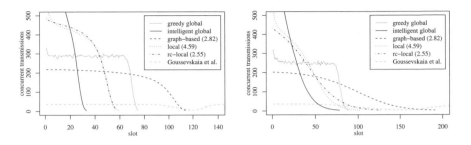

Fig. 6. Utilization of slots for different scheduling algorithms, for 20000 randim links (left) and random UDGs with ≈20000 links, averaged over 250 runs. Plotted are results for the graph-based $\Omega(1)$-sender models $(2.82\ell(\mathcal{M}_G), 2.82\ell(\mathcal{M}_G))$ and $(4.59\ell(\mathcal{M}_G), \ell(\mathcal{M}_G))$ and the model $(2.55\ell(\mathcal{M}_G), 1.75\ell(\mathcal{M}_G))$ for the receiver-centered locality. Values for ρ and c are minimal in the sense that by decreasing these parameters we cannot prove correctness with respect to $\mathcal{M}_G$. Results are compared to the following global algorithms : greedy scheduling, intelligent scheduling and the algorithm from [8].

graph-based, but allows for better bounds and schedules. Table 1 shows values for random links, averaged over 250 runs, a selection is also plotted in Fig. 6 together with results from scheduling UDG links. Not surprisingly, the more far-seeing global algorithm performs best among all compared schemes and the global algorithm from [8] by far worst. Among the greedy schedule algorithms, the global view did in general not give an advantage. Greedy scheduling in the SINR_G model only outperformed the graph-based variant (whose big advantage is its simplicity). Increasing locality a little more or switching to the receiver-centered locality, the local models even led to better results since the exclusion radius prevented scheduling of close links and receiver-centered locality reflects the nature of interference better.

7   Conclusion and Future Work

In this work, we introduced the concept of local interference models capturing the natural demand for scalable and distributed scheduling protocols to have a local yet provably correct characterization of successful concurrent transmissions. We prove lower bounds that arise in those models and introduce a very simple class, $\Omega(1)$-sender models which provably allow for constant approximation compared to any other local interference model that produces correct results with respect to the widely agreed on geometric SINR model. We believe that these models will be helpful to attack other local problems such as the problem to construct good topologies not only with respect to properties of spanning ratios and low degree or other heuristics to minimize interference, but with respect to the local construction of short schedules of these topologies. To this extent, it will also be of interest to generalize the findings of this work to the case of variable power assignment, which, unfortunately is not well understood even with respect to global schedule algorithms.

References

1. Schmidt, S., Wattenhofer, R.: Algorithmic Models for Sensor Networks. In: 20th IEEE Int. Parallel and Distributed Processing Symposium (IPDPS 2006), pp. 450–459 (2007)
2. Rappaport, T.: Wireless Communications: Principles and Practices. Prentice-Hall, Englewood Cliffs (1996)
3. Gupta, P., Kumar, P.R.: The Capacity of Wireless Networks. IEEE Transactions on Information Theory 46(2), 388–404 (2000)
4. Moscibroda, T., Wattenhofer, R., Weber, Y.: Protocol Design Beyond Graph-Based Models. In: Proc. of the 5th Workshop on Hot Topics in Networks (HotNets) (2006)
5. Hunt, H., Marathe, M., Radhakrishnan, V., Ravi, S., Rosenkrantz, D., Stearns, R.: NC-Approximation Schemes for NP- and PSPACE-Hard Problems for Geometric Graphs. Journal of Algorithms 26 (1998)
6. Krumke, S.O., Marathe, M., Ravi, S.S.: Models and approximation algorithms for channel assignment in radio networks. Wireless Networks 6, 575–584 (2000)
7. Björklund, P., Värbrand, P., Yuan, D.: A Column Generation Method for Spatial TDMA Scheduling in Ad hoc Networks. Ad Hoc Networks 2(4), 4005–4418 (2004)
8. Goussevskaia, O., Oswald, Y.A., Wattenhofer, R.: Complexity in Geometric SINR. In: Proceedings of the 8th ACM International Symposium on Mobile Ad Hoc Networking and Computing (MOBIHOC 2007), pp. 100–109. ACM Press, New York (2007)
9. Locher, T., von Rickenbach, P., Wattenhofer, R.: Sensor Networks Continue to Puzzle: Selected Open Problems. In: Rao, S., Chatterjee, M., Jayanti, P., Murthy, C.S.R., Saha, S.K. (eds.) ICDCN 2008. LNCS, vol. 4904, pp. 25–38. Springer, Heidelberg (2008)
10. Hajek, B., Sasaki, G.: Link Scheduling in Polynomial Time. IEEE Transactions on Information Theory 34(5), 910–917 (1988)
11. Kumar, V.S.A., Marathe, M.V., Parthasarathy, S., Srinivasan, A.: End-to-end packet scheduling in wireless ad-hoc networks. In: Proc. of the 15th annual ACM-SIAM symposium on Discrete Algorithms (SODA 2004), pp. 1021–1030 (2004)
12. Moscibroda, T., Wattenhofer, R.: Coloring Unstructured Radio Networks. In: Proc. of the 17th Annual ACM Symp. on Parallel Algorithms and Architectures (SPAA 2005) (2005)
13. Behzad, A., Rubin, I.: On the Performance of Graph-based Scheduling Algorithms for Packet Radio Networks. In: Proc. of the IEEE Global Telecommunications Conference (GLOBECOM) (2003)
14. Grönkvist, J., Hansson, A.: Comparison Between Graph-Based and Interference-Based STDMA Scheduling. In: Proc. of the 2nd ACM International Symposium on Mobile Ad Hoc Networking & Computing (MOBIHOC), pp. 255–258 (2001)
15. Moscibroda, T., Wattenhofer, R.: The Complexity of Connectivity in Wireless Networks. In: Proceedings of the 25th Annual Joint Conference of the IEEE Computer and Communications Societies (INFOCOM 2006) (2006)
16. Moscibroda, T., Oswald, Y.A., Wattenhofer, R.: How Optimal are Wireless Scheduling Protocols? In: Proceedings of the 26th Annual Joint Conference of the IEEE Computer and Communications Societies (INFOCOM 2007) (2007)

17. Linial, N.: Locality in Distributed Graph Algorithms. SIAM Journal on Computing 21, 193–201 (1992)
18. Naor, M., Stockmeyer, L.: What can be computed locally? SIAM Journal on Computing 24(6), 1259–1277 (1995)
19. Kuhn, F., Moscibroda, T., Wattenhofer, R.: The Price of Being Near-Sighted. In: Proc. of the 17th ACM–SIAM Symp. on Discrete Algorithms (SODA 2006) (2006)
20. Toth, L.F.: Minkowskian distribution of discs. Proceedings of the AMS 16(5), 999–1004 (1965)

Algorithms for Location Estimation
Based on RSSI Sampling

Charalampos Papamanthou, Franco P. Preparata, and Roberto Tamassia

Department of Computer Science and Center for Geometric Computing
Brown University
{cpap,franco,rt}@cs.brown.edu

Abstract. In this paper, we re-examine the RSSI measurement model
for location estimation and provide the first detailed formulation of the
probability distribution of the position of a sensor node. We also show
how to use this probabilistic model to efficiently compute a good estimation of the position of the sensor node by sampling multiple readings
from the beacons (where we do not merely use the mean of the samples)
and then minimizing a function with an acceptable computational effort.
The results of the simulation of our method in TOSSIM indicate that
the location of the sensor node can be computed in a small amount of
time and that the quality of the solution is competitive with previous
approaches.

1 Introduction

Estimating the location of a roaming sensor is a fundamental task for most
sensor networks applications. For example, if a sensor network has been deployed
to provide protection against fire (in this case, sensor nodes report a sudden
increase in temperature), we want to know the location of the sensor that triggers
an alert so that action can be taken accordingly. Additionally, some routing
protocols for sensor networks, such as geographical routing [15, 44], make routing
decisions based on the knowledge of the locations of the sensor nodes. Common
location estimation protocols that are widely adopted in practice assume that
there are some fixed nodes (base stations) that know their location which are
called *beacons*. These nodes send a signal to the sensor nodes that want to
determine their location. According to the intensity (or for example the angle)
of this signal, the sensor node can have an estimate of the distance between them
and the beacons.

After performing a certain number of such measurements for different beacons, the sensor node has to combine all this information (for RSSI (Received
Signal Strength Indicator), this information is the power of each individual signal and the coordinates of the corresponding transmitter) in order to estimate
its location. However one could ask the following question: Why cannot we use a
Geographic Positioning System (GPS) to efficiently achieve the task of localization? The answer is that a GPS requires a strong computing platform which is
not available in sensor networks. Sensor nodes are typically very low-computing

S. Fekete (Ed.): ALGOSENSORS 2008, LNCS 5389, pp. 72–86, 2008.

power units that can efficiently perform only basic arithmetic operations; Requiring the execution of complex arithmetic operations on a sensor node would entail a quick depletion of its battery which is not desirable for most practical applications. Finally, the localization problem gets even more difficult because the available power on the sensor node is limited: therefore no accurate measurements of the signal can be made (since an accurate measurement requires more computing power) which means that the measurements are prone to errors. This is something that should also be taken into consideration and treated accordingly.

Therefore, any location estimation algorithm should have the following requirements:

1. The sensor node should avoid complex and time consuming computations, which would deplete its energy supply (typically a low-cost battery) rapidly;
2. The computations should take into consideration the error in the measurements, which can be large.

1.1   Related Work and Observations

There are several proposed location estimation protocols for sensor networks, see, e.g., [5, 7, 8, 11, 23, 24, 25, 26, 27]. All these protocols use the same model, where some nodes *know* their location (either because they are fixed or by using GPS) and are called *beacons* or *anchor nodes*, and some other nodes, called *sensor nodes*, estimate their location using the information they receive from the beacons. This information consists of the beacons' coordinates and of features of the beacon signal, such as the *received signal strength indicator* (RSSI) or the *time difference of arrival* (TDoA). Also, other protocols (e.g., [26]) are based on the capability of the nodes to sense the angle from which a signal is received. Recently [38] presented a solution for aerial localization and [22] proposed a solution where the localization is based on adopting slightly different periodic signal frequencies (interferometric positioning). This solution [22] is very competitive and achieves very good precision. However the computations used are very intensive.

Several previous approaches use computationally demanding methods, such as convex optimization [8], systems of complex equations [26], *minimum mean square error* (*MMSE*) methods [6, 33], and Kalman filters [34]. In these approaches, the measurement model is not adequately analyzed and the error is assumed to be small, which is not the case in most real applications of sensor networks.

Other approaches, notably [17, 30, 36], estimate the location of a node using the RSSI method (analyzed in [31]), which is the most realistic model for sensor network communication. In [17], the authors evaluate the ability of a sensor network to estimate the location of sensor nodes. They assume that the location of the sensor node is known and develop arguments concerning the probability that the network will detect this location. They use the RSSI error model to analyze the problem of *evaluating* the ability of the sensor network to locate a sensor node. However, they do not describe how their algorithms can be implemented on a sensor node to estimate its own location. Moreover, their method does not take into account the basic parameters of the RSSI model (standard deviation and path loss exponent) and thus gives incorrect results.

In this paper, we formulate the correct probability distribution of the position of a sensor node based on one reading produced with the RSSI model. Due to the errors implicit in the RSSI model, it is unrealistic to try to compute a good estimation of the location of the sensor node based only on a single measurement (or even few measurements) from each beacon. Such an approach would be so inaccurate to make the estimate practically worthless. Notwithstanding this difficulty, we show that a reliable estimation of the location can be achieved by processing a reasonably small number of readings of the signals.

Especially for indoor positioning systems, this is an assumption that has been extensively used. For example, in [13, 14], the position estimation is based on a *location fingerprint* $\mathbf{t} = [t_1 \ t_2 \ldots t_N]$, where N is the number of beacons and t_i $(i = 1, \ldots, N)$ is the mean value of the received signal strength over a certain time window. Also, in [3, 9, 20, 42], experiments with various sample sizes are presented where the samples are used to compute certain features of the signal strength such as the standard deviation and the path loss exponent. Finally, [12] presents simulations that use various number of samples, where more than 50 samples to *filter out* the errors in the probability distribution are used.

We show that using only the mean of the measurements is not a correct procedure, due to the lognormal distribution of the distance from the beacon (see Theorem 3). Instead of using directly the mean value, we use another value that is adequate according to the specific underlying probability distribution of the distance. The number of samples that are used vary from 20 to 60 and obviously the accuracy of the computed location grows with the number of samples. Finally, once the sampling has been performed, we show how to seek the minimum of a function that approximates the actual location with small computational effort.

1.2 Our Contributions

The main contributions of this paper are as follows:

1. We evaluate the probability that a sensor node lies within a certain region, given that the power received from the beacons is modeled with RSSI. To the best of our knowledge, this is the first detailed formulation of the probability distribution of the position of a sensor node. We show that unlike the normal distribution of the received power, the probability distribution of the actual position is lognormal. Thus, we give evidence to the role of the parameters σ and n in the probability distribution of the actual distance, where σ is the standard deviation of the normal variable that models the power received by the sensor and n is a parameter, called *path loss exponent*, that depends on the transmission medium. In previous approaches [17], the probability distributions used did not exhibit dependency on these two variables.

2. We present a method for estimating the location of a node from multiple sample power readings from the beacons. Our method computes the *expected*

value of the received power and combines it with the mean and the standard deviation of the sample readings using a steepest descent approach [37]. We show that our method is simple and efficient and provides a good estimation of the position. Note that using multiple sample readings is necessary for a reliable location estimation since the probability distribution of the location for a single sample implies that the domain within which the sensor lies with high probability has large area.

3. We describe an implementation of our location estimation algorithm that is suitable for execution on standard sensor hardware and we analyze the results of an extensive simulation of the execution of the algorithm in TOSSIM [10, 18]. Our simulation shows that our method has accuracy that is comparable to or better than that of previous methods.

1.3   Organization of the Paper

The rest of this paper is organized as follows. In Section 2, we overview the RSSI model, give formulas for the probability distribution of the position of a sensor node due to power measurements, and show how to estimate the actual distance given a set of sample power readings. We develop an efficient algorithm for location estimation and analyze its running time in Section 3. Finally, in Section 4, we report the results of the simulation and present a comparison (in terms of localization error) of our method with previous approaches. Concluding remarks are in Section 5.

2   Theoretical Framework

This section provides a formal probabilistic framework for estimation of the position of a sensor node from power measurements.

2.1   RSSI Model

Suppose we are given a region of the plane with k beacon nodes $b_1, b_2, \ldots, b_k$ (nodes of known location). The coordinates of the beacons are (x_i, y_i) for $i = 1, \ldots, k$. The beacons transmit information about their location with a signal of normalized intensity to a sensor node s that does not know its location. Based on the locations of the beacons and the estimated distances from the beacons (computed from the received signals), the sensor is to compute its actual location.

Among the several models proposed for estimating the distance between a beacon and a sensor node, the most realistic and commonly used one is the *received signal strength indicator* model (RSSI) [31]. In this model, the beacon broadcasts signal to all sensors and the sensors can estimate the distance between them and the beacons on the basis of the strength of the signals they receive.

Let b_i be a beacon located at (x_i, y_i) and s a sensor node located at (x, y). We define the *relative error* ϵ_i pertaining to b_i as follows. Suppose that s reads a distance $\hat{r}_i$, while the actual distance is r_i. The relative error is

$$\epsilon_i = \frac{\hat{r}_i}{r_i} - 1 \in [-1, +\infty). \tag{1}$$

The commonly accepted transmission model [31] expresses the received power p_i (in dBm) as

$$p_i = p_0 + 10n \log\left(\frac{r_i}{r_0}\right) \tag{2}$$

where p_0 is the received power in dBm at a reference distance r_0 and n is the path loss exponent which is a constant depending on the transmission medium (indoors, outdoors) and ranges typically from 2 to 4. In some environments, such as buildings, stadiums and other indoor environments, the path loss exponent can reach values in the range of 4 to 6. On the other hand, a waveguide type of propagation may occur in tunnels, where the path loss exponent drops below 2.

We recall that if the received power in mW at a point k is P_k, and $P_{k'}$ is the received power at some reference point k' (again in mW), then the received power p_k in dBm at point k is defined as

$$p_k = 10 \log\left(P_k/P_{k'}\right). \tag{3}$$

The measured power, however, differs from that given in Equation (2); due to channel fading (variation of the received signal power caused by changes in transmission medium or path), the measured power is $\hat{p}_i = p_i + x$. The random variable x represents the medium-scale channel fading and is typically modelled as Gaussian zero-mean with variance σ^2 (in dBm). Typically, σ is as low as 4 and as high as 12 (this implies that the error may be large). Inserting $\hat{p}_i$ and $\hat{r}_i$ into (2), we get

$$\hat{p}_i = p_0 + 10n \log\left(\frac{\hat{r}_i}{r_0}\right) \tag{4}$$

where now the measured power $\hat{p}_i$ in dBm relates to the measured distance $\hat{r}_i$ by the sensor. By combining the above equations, we get that the relation between the measured distance and the actual distance is

$$\hat{r}_i = r_i 10^{\frac{x}{10n}} \tag{5}$$

which gives

$$\epsilon_i = 10^{\frac{x}{10n}} - 1. \tag{6}$$

2.2   Probability Distributions

In this section, we present the probability distribution of the position of the sensor node based on measurements of one or more beacons. We denote e^x with $\exp(x)$.

Theorem 1. *Let b_i be a beacon node located at (x_i, y_i) sending information to a sensor node under the RSSI model with standard deviation σ and path loss exponent n. Let $\hat{r}_i$ be the measured distance from beacon node b_i at the sensor*

node. We have that the probability density function of the actual position (x, y)
of the sensor node is given by

$$P_{X,Y}^{(i)}(x, y) = \frac{10n \exp\left(-\left(10n \log \frac{\hat{r}_i}{\sqrt{(x-x_i)^2+(y-y_i)^2}}\right)^2 / 2\sigma^2\right)}{2\pi\sigma\sqrt{2\pi}\ln(10)((x-x_i)^2+(y-y_i)^2)}.$$

To simplify the notation, we denote the probability distribution due to beacon
b_i with

$$\Phi_{b_i}(x, y) = P_{X,Y}^{(i)}(x, y).$$

The previous argument can be extended to a finite set of beacons $B = \{b_1, \ldots, b_k\}$ yielding the following theorem:

Theorem 2. *Let* $B = \{b_1, b_2, \ldots, b_k\}$ *be a set of beacons sending information to a sensor node under the RSSI model with standard deviation* σ *and path loss exponent* n. *If the measured distance from beacon node* b_i *at the sensor node is* $\hat{r}_i$ *(*$i = 1, \ldots, k$*), then the probability density function (due to all the beacons in B) of the actual position* (x, y) *of the sensor node is given by*

$$\Phi_{(B)}(x, y) = \frac{\prod_{i=1}^{k} \Phi_{b_i}(x, y)}{\int_{-\infty}^{+\infty} \int_{-\infty}^{+\infty} \left(\prod_{i=1}^{k} \Phi_{b_i}(x, y)\right) dy dx}$$

where $\Phi_{b_i}(x, y)$ *is the probability distribution due to beacon* b_i, *as defined in Theorem 1.*

As we will see later, if we use only one measurement, we may end up with a density function having more than one maximum. Moreover, the computation of this maximum on a sensor node is a difficult task since there is no simple analytical expression for the maximum of the probability distribution. Also, there is no suitable analytical expression for the integrals needed to compute the probability for a certain region. Due to space limitations, the proofs of the two theorems can be found in the full version of the paper.

2.3   Samples of Measurements

In the previous sections, we have examined the probability distribution of the sensor's position based on a single measurement. This setting, however, can give rise to unacceptable errors for the values of σ ($\sigma = 4, 6, 8$) reported in the literature [31]. A consequence of this situation is that we may be unable to define a disk containing the sensor's location with an acceptable degree γ of confidence (say, $\gamma > 0.9$). Additionally, if our practice is based on only one reading, there is no way for the sensor to estimate the standard deviation σ (this is actually the standard deviation of the Gaussian random variable x introduced in Section 2.1) of each beacon. This parameter σ—conveniently assumed to be known in

Section 2.1—must be estimated in practice as it is needed in the computations (see below).

To overcome these difficulties, we show in this section how we can obtain a good estimate of the location of the sensor node based on small number of readings. Especially for indoor positioning systems, this is an assumption that has been extensively used. For example, in [13, 14], the position estimation is based on a *location fingerprint* $\mathbf{t} = [t_1 \, t_2 \ldots t_N]$, where N is the number of beacons and t_i $(i = 1, \ldots, N)$ is the mean value of the received signal strength from the i-th beacon over a certain time window. Note that t_i denotes the *measured* power $\hat{p}_i$ that appears in Equation 4. Hence, the "measured power" location fingerprint $\mathbf{t}$ can be transformed to a location fingerprint $\mathbf{r}$ of "measured radii" by using Equation 4, since the reference values p_0 and d_0 are known. The number of samples that are used vary from 20 to 60 and obviously this number affects the accuracy of the computed location. Also, in [3, 9, 20, 42], experiments with various sample sizes are presented where the samples are used to compute certain features of the signal strength such as the standard deviation and the path loss exponent.

Suppose now that we use a sample of k readings from beacon b_i. We have a sequence of radii $\hat{r_{i1}}, \hat{r_{i2}}, \ldots, \hat{r_{ik}}$. Let $\bar{r}_i$, $\bar{s}_i$ denote the *unbiased* estimators of the value $\mathsf{E}[\hat{r}_i]$ and of the standard deviation $\sqrt{\mathsf{Var}(\hat{r}_i)}$ of the underlying distribution of the measured radii $\hat{r}_i$ $(i = 1, \ldots, 3)$ respectively. We have the following result that relates estimates of the actual distance and the standard deviation with reference to a beacon b_i with features of the lognormal distribution:

Theorem 3. *Suppose a sensor node reads k distance samples $\hat{r_{i1}}, \hat{r_{i2}}, \ldots, \hat{r_{ik}}$ from a beacon b_i that is modeled with the RSSI of path loss exponent n and standard deviation σ_i. If $\bar{r}_i$ is the sample mean and $\bar{s}_i$ is the sample standard deviation then we have the following:*

1. *The estimate of the square of the actual distance r_i^2 from beacon b_i is*

$$\frac{\bar{r}_i^{\,4}}{\bar{r}_i^{\,2} + \bar{s}_i^{\,2}}.$$

2. *The estimate of the square of the standard deviation σ_i^2 is*

$$\frac{100n^2}{\ln^2(10)} \ln \left[1 + \left(\frac{\bar{s}_i}{\bar{r}_i} \right)^2 \right].$$

Note that the above theorem indicates that the quality of estimation of the actual distance is heavily dependent on the estimation of the distribution of the measured radii.

3   Location Estimation

In this section, we develop an algorithm for location estimation based on several samples. This algorithm does not involve any complex calculations (such as square roots) which is very important to consider when we develop algorithms to be executed on sensor nodes, due to the sensor's modest computing power.

3.1   Algorithm

As we saw in the previous section, after completing the sampling procedure, we derive estimates for r_1^2, r_2^2, r_3^2, given by Theorem 3. Our aim is to formulate a function whose minimum will yield a good approximation of the sensor's location. This function should be convex and also its derivatives should not include roots. Suppose now we have three beacons located at (x_1, y_1), (x_2, y_2), (x_3, y_3). Let $f(x, y)$ be the function

$$f(x, y) = \sum_{i=1}^{3} ((x - x_i)^2 + (y - y_i)^2 - r_i^2)^2. \tag{7}$$

Note that if all 3 circles intersect at the same point (x_0, y_0), this function has minimum 0 at (x_0, y_0). Unfortunately, minimizing that function is not an easy task, if we are restricted on the available primitives. Hence we are going to use methods that are based on the gradient of the function. The good feature about such methods is that we can get to a point very close to the minimum in a small number of computationally simple iterations. Indeed, let

$$\alpha(x, y) = \frac{\partial f(x, y)}{\partial x} = 4 \sum_{i=1}^{3} (x - x_i)((x - x_i)^2 + (y - y_i)^2 - r_i^2) \tag{8}$$

and

$$\beta(x, y) = \frac{\partial f(x, y)}{\partial y} = 4 \sum_{i=1}^{3} (y - y_i)((x - x_i)^2 + (y - y_i)^2 - r_i^2) \tag{9}$$

be the partial derivatives of f. Note that the above expressions are computable on a sensor node. The function $z = f(x, y)$ describes a convex solid surface with obvious definitions of "interior" and "exterior". Initially, we make a guess for our point (this is required by all steepest descent methods [37]). Suppose, for uniformity, we choose as our initial point (x_0, y_0) the centroid of the beacon triangle. We compute the vector v which is orthogonal to the tangent plane $\mathcal{T}$ and pointing toward the exterior. Hence $\mathsf{v} = \begin{bmatrix} \alpha(x_0, y_0) & \beta(x_0, y_0) & -1 \end{bmatrix}^T$. Let now $\mathcal{P}$ be the *vertical* plane containing v applied to $(x_0, y_0, f(x_0, y_0))$. Since $\mathcal{P}$ is a vertical plane, any normal vector w of $\mathcal{P}$ will have a zero z-component. Additionally, w is orthogonal to v and therefore may be chosen as $\mathsf{w} = \begin{bmatrix} -\beta(x_0, y_0) & \alpha(x_0, y_0) & 0 \end{bmatrix}^T$. We seek the vector q pointing towards the minimum of the function. Such vector belongs to $\mathcal{P}$ and is orthogonal to v (q is orthogonal both to v and w), i.e.,

$$\mathsf{q} = \begin{bmatrix} \alpha(x_0, y_0) & \beta(x_0, y_0) & \alpha^2(x_0, y_0) + \beta^2(x_0, y_0) \end{bmatrix}^T.$$

Now we compute the intersection point (x_0', y_0') of the line (with the surface $z = 0$) passing by $(x_0, y_0, f(x_0, y_0))$ which is collinear with the direction q and the xy-plane. The parametric equation of this line is $(x, y, z) = (x_0 + t\mathsf{q}_x, y_0 + t\mathsf{q}_y, f(x_0, y_0) + t\mathsf{q}_z)$ for all $t \in \mathbf{R}$. The new point (x_0', y_0') is then given by

$$(x_0', y_0') = \left(x_0 - \frac{f(x_0, y_0)\alpha(x_0, y_0)}{\alpha^2(x_0, y_0) + \beta^2(x_0, y_0)}, y_0 - \frac{f(x_0, y_0)\beta(x_0, y_0)}{\alpha^2(x_0, y_0) + \beta^2(x_0, y_0)} \right).$$

The described process gives a new point (x'_0, y'_0). This point is expectedly closer to the point that corresponds to the minimum of f as we follow the direction of the gradient as long as the products $\alpha(x_0, y_0)\alpha(x'_0, y'_0) > 0$ and $\beta(x_0, y_0)$ $\beta(x'_0, y'_0) > 0$. When this condition no longer holds, we have "overshot"; to remedy, we backtrack to the previous point referred here as (x, y) and apply a typical steepest descent method with very small rate λ. We therefore compute our new point (x', y') by setting

$$(x', y') = (x - \lambda\alpha(x, y), y - \lambda\beta(x, y)). \tag{10}$$

We continue this process until the gradients $\alpha(x, y)$, $\beta(x, y)$ change sign. At that point we stop and we report the final point as our estimation. Here we should emphasize the fact that it is important to take samples of adequate size. Taking samples implies a better behavior for f, meaning that there would be only one minimum and therefore the algorithm will quickly converge to the minimum. As far as the value of the variable λ is concerned, this variable is chosen to be small enough and inversely proportional to the size of the grid since these features of λ force the second **repeat** loop of the algorithm to converge quickly. This has been observed in the experiments. For the experiments, the value of λ is equal to $1000^{-m/100}$.

3.2   Complexity and Limitations

The most expensive part of the presented algorithm is the sampling procedure and the computation of the estimates $\bar{r}_i^2$ and $\bar{s}_i^2$. These steps take time $O(k)$, where k is the number of samples. There is an obvious trade-off between accuracy and power consumption. Also, the computation executed on the sensor nodes depends on the time the gradient methods take to converge, which is generally small for a well-behaved function. For the other parts of the algorithm there are closed formulas, so we can assume that they take time $O(1)$. We can also see that the exact number of multiplications needed by the presented program is $(7k + 8) + 10n_1 + 4n_2$, where is k is the size of the sample, n_1 is the number of iterations of the first **repeat** loop and n_2 is the number of iterations of the second **repeat** loop.

The size of the code of the program (ROM) written in NesC [10] is 47K whereas the amount of memory (RAM) needed to execute this program in TOSSIM [18] (see Section 4) is 637K. As far as the complexity of the closed formulas computation is concerned it is realistic to assume that the involved in closed-formula calculation can be executed on a sensor node (essentially floating point operations). For example, there are micro-controllers, such as the ATM-Mega128L [2] and MSP430 [39], which have very rich instruction sets. Finally, a hardware multiplier allows floating-point arithmetic to be carried out [21].

4   Simulation

In this section, we present and analyze extensive simulation results of our method. We have run our experiments with TOSSIM [10, 18], a widely used simulator of the TinyOS operating system for sensor networks.

We executed our simulations in a square of area $m \times m$ cells, where $m = 50, 100, 200$. The three beacons are placed in positions that form a well conditioned triangle (well-conditioning is synonymous with the fact that the function $f(x, y)$ has a single global minimum). Namely, the first beacon is placed at $(0, 0)$, the second beacon is placed at $(m, 0)$ and the third beacon is placed at $(m/2, 3m/4)$. The standard deviations of the three beacons $\sigma_1, \sigma_2, \sigma_3$ are set to 4 and the path loss exponent n is set to 2. We also recall that we set the variable λ that appears in Equation 10 equal to $1000^{-\frac{m}{100}}$, where m is the dimension of the grid. Finally the measured distance is computed using Equation 5.

We measure the execution time of the algorithm implemented in NesC [10] (that runs in TinyOS) and the average number of iterations of the repeat loops over 1000 runs. We also determine the mean of the distance d between the actual point and the computed point. We use the ratio $\frac{d}{m}$ to evaluate the quality of the solution computed by the algorithm. Note that this metric was proposed in [41].

The results obtained for different numbers of samples and different sizes of grids are shown in Table 1 , where m is the dimension of the square region, k is the number of samples, the time is counted in milliseconds (we count the exact time that the simulated processor in TOSSIM takes to execute this program), n_1 is the number of iterations of the first repeat loop, n_2 is the number of iterations of the second repeat loop and d is the mean of the distance between the actual point and the computed point.

The simulation results with TOSSIM show that the sensor node can execute the algorithm in a small amount of time. This time is proportional to the number of samples we use each time, which indicates that the sampling procedure dominates the execution time on the sensor node. Only up to six iterations $(n_1 + n_2)$ are enough to compute an estimation of the actual point and the quality of the estimation is dependent on the number of the samples. Additionally, note that for various grid sizes, the algorithm has a uniform behavior since the ratio $\frac{d}{m}$ is similar for different sizes of the grid. Finally, in all cases, the solution we get is better for larger sizes of samples. For example, though not practical for power consumption issues, for 200 samples the estimation gets even better ($d/m = 0.041$ for $m = 200$).

Table 1. Results of the simulation in TOSSIM for an $m \times m$ square grid ($m = 50, 100, 200$): execution time, average number of iterations of the program (n_1, n_2), localization error (d) and ratio $\frac{d}{m}$ for various samples sizes over 1000 runs

$m \times m$	k	time (ms)	n_1	n_2	d	d/m
50×50	20	0.140	4.045	1.081	5.018	0.1003
50×50	40	0.230	4.226	1.061	3.774	0.0745
100×100	20	0.130	3.670	1.142	9.986	0.0986
100×100	40	0.240	3.817	1.046	7.634	0.0763
200×200	20	0.120	3.640	2.736	19.977	0.0998
200×200	40	0.240	3.820	2.323	14.957	0.0747

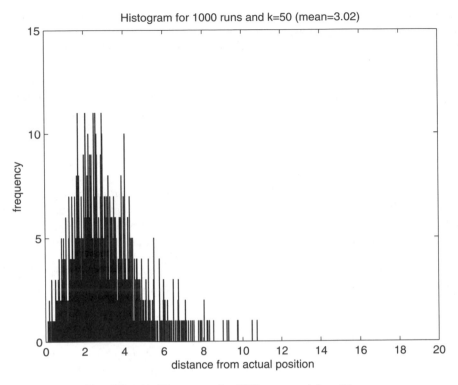

Fig. 1. Histogram for 1000 runs and $k = 50$

We show in Figure 1 the probability distribution of the distance of the computed point from the actual point derived for 1000 runs and for $m = 50$. In particular, we show the plot of the distribution of the error of the estimation (which we define as the distance of the computed point from the actual point) for sample size $k = 50$. The mean of the error for these measurements is about 3 for $k = 50$. Also, we observe that the distribution of the error appears to follow a lognormal distribution.

Finally, Table 2 compares location estimation algorithms. We present the average localization error d, the area A of the field where the experiments are executed and the ratio $\frac{d}{\sqrt{A}}$. We use as a comparison measure the quantity $\frac{d}{\sqrt{A}}$, which for our algorithm is bounded by 0.1 (this is what we get for the smallest number of samples $k = 20$). However, for a number of samples $k = 300$ (though not so practical) we can get even smaller values (for example for $k = 300$ and for an area 1000×1000 we get $\frac{d}{\sqrt{A}} = 0.026$).

From Table 2, we see that our method gives always better or as good results as the results obtained by the existing methods (except for [38] where, however, the time needed for localization ranges from 10 milliseconds to 2 minutes — something that is also observed in [22]— where the precision is even better). Also, if we slightly increase the number of the samples we use, we get very good results and the ratio drops substantially (for example for $k = 60$ we get a ratio

Table 2. Comparison of existing work. In each row, we display the bibliographic reference and the respective average localization error (d), the size of the area of the experiments A, the ratio $\frac{d}{\sqrt{A}}$ and finally the number of samples used by each method. Note that it is not always feasible to compare between different methods since the settings used can be different. N.A. stands for "not applicable" and it means that the certain method does not refer explicitly to the number of samples used or that the sampling technique is not used.

reference	error d	simulation area A	$d/\sqrt{A}$	number of samples
[3]	3	22.5×45.5	0.090	20
[32]	3	16×40	0.118	N.A.
[4]	4	35×40	0.107	250
[29]	7.62	13.71×32	0.360	40
[1]	3	500	0.130	N.A.
[30]	6	60×60	0.100	N.A.
[5]	1.83	10×10	0.183	20
[35]	0.8	6×6	0.130	50
[19]	13	18751	0.094	N.A.
[43]	10	26×49	0.280	N.A.
[38]	3.5	60×120	0.058	N.A.
[16]	0.82	5×5	0.164	N.A.
our scheme	4.350	50×50	$\simeq 0.087$	25
our scheme	3.020	50×50	$\simeq 0.064$	50

$\frac{d}{\sqrt{A}} = 0.06$). Note that previous methods use more than three beacon nodes (see for example [28] where $O(m)$ beacons are placed in the area of localization for an $m \times m$ grid).

5   Conclusions

In this paper, we have analyzed the RSSI model for location estimation in sensor networks. Given a normal distribution for the error in dBm, we compute the correct probability distribution of the sensor's location and then we adopt this probability distribution in a theoretical analysis of sampling the measurements for location estimation. We finally give a simple algorithm that can be executed on sensor nodes; its complexity, for a constant number of beacons, is proportional to the size of the sample.

Location estimation in sensor networks presents several trade-offs. If higher accuracy is desired, one has to deploy more beacons or use more samples. Using a large number of beacons and samples causes significant energy consumption. The energy-optimal case occurs when only three beacons are deployed and an estimation of the actual point is based on the probability distribution computed by taking into consideration only one measurement. This solution, however, gives unacceptable errors. Additionally, performing computations with the exact probability distribution is unrealistic, since it involves complex formulas. Hence, were we to depend on few measurements, off-line computed data must be stored as tables within the sensor, which immediately creates a storage problem. However, one can use more samples, thus increasing energy consumption.

Acknowledgments

This research was supported by the U.S. National Science Foundation under grants IIS–0324846 and CCF–0830149 and by the Center for Geometric Computing and the Kanellakis Fellowship at Brown University. The views in this paper do not necessarily reflect the views of the sponsors. We thank Goce Trajcevski for useful discussions.

References

[1] Alippi, C., Vanini, G.: A RSSI-based and calibrated centralized localization technique for wireless sensor networks. In: Proc. IEEE Int. Conf. on Pervasive Computing and Communications Workshops (PERCOMW), pp. 301–306 (2006)

[2] Atmel Corporation. ATM128 Datasheet, Revised 2461-09/03 (2003)

[3] Bahl, P., Padmanabhan, V.N.: RADAR: An in-building RF-based user location and tracking system. In: Proc. IEEE Conf. on Computer Communications (INFOCOM), pp. 775–784 (2000)

[4] Brunato, M., Battiti, R.: Statistical learning theory for location fingerprinting in wireless LANs. Computer Networks 47(6), 825–845 (2005)

[5] Bulusu, N., Heidemann, J., Estrin, D.: GPS-less low cost outdoor localization for very small devices. IEEE Personal Communications Magazine 7(5), 28–34 (2000)

[6] Capkun, S., Hubaux, J.-P.: Secure positioning of wireless devices with application to sensor networks. In: Proc. IEEE Conf. on Computer Communications (INFOCOM), pp. 1917–1928 (2005)

[7] Dil, B., Dulman, S., Havinga, P.: Range-based localization in mobile sensor networks. In: Römer, K., Karl, H., Mattern, F. (eds.) EWSN 2006. LNCS, vol. 3868, pp. 164–179. Springer, Heidelberg (2006)

[8] Doherty, L., Pister, K.S.J., Ghaoui, L.E.: Convex optimization methods for sensor node position estimation. In: Proc. IEEE Conf. on Computer Communications (INFOCOM), pp. 1655–1663 (2001)

[9] Faria, D.B.: Modeling signal attenuation in IEEE 802.11 wireless LANs. vol. 1. Technical Report TR-KP06-0118, Stanford University (2005)

[10] Gay, D., Levis, P., von Behren, R., Welsh, M., Brewer, E., Culler, D.: The nesC language: A holistic approach to networked embedded systems. In: Proc. ACM Conf. on Programming Language Design and Implementation (PLDI), pp. 1–11 (2003)

[11] He, T., Huang, C., Blum, B.M., Stankovic, J.A., Abdelzaher, T.: Range-free localization schemes for large scale sensor networks. In: Proc. of the Int. Conf. on Mobile Computing and Networking (MOBICOM), pp. 81–95 (2003)

[12] Hu, L., Evans, D.: Localization for mobile sensor networks. In: Proc. of the Int. Conf. on Mobile Computing and Networking (MOBICOM), pp. 45–57 (2004)

[13] Kaemarungsi, K., Krishnamurthy, P.: Modeling of indoor positioning systems based on location fingerprinting. In: Proc. IEEE Conf. on Computer Communications (INFOCOM), pp. 1012–1022 (2004)

[14] Kaemarungsi, K., Krishnamurthy, P.: Properties of indoor received signal strength for WLAN location fingerprinting. In: Proc. Int. Conf. on Mobile and Ubiquitous Systems (MOBIQUITOUS), pp. 14–23 (2004)

[15] Karp, B., Kung, H.T.: GPSR: Greedy perimeter stateless routing for wireless networks. In: Proc. of the Int. Conf. on Mobile Computing and Networking (MO-BICOM), pp. 243–254 (2000)

[16] Krohn, A., Hazas, M., Beigl, M.: Removing systematic error in node localisation using scalable data fusion. In: Langendoen, K.G., Voigt, T. (eds.) EWSN 2007. LNCS, vol. 4373, pp. 341–356. Springer, Heidelberg (2007)

[17] Kuo, S.-P., Tseng, Y.-C., Wu, F.-J., Lin, C.-Y.: A probabilistic signal-strength-based evaluation methodology for sensor network deployment. In: Proc. Int. Conf. on Advanced Information Networking and Applications (AINA), pp. 319–324 (2005)

[18] Levis, P., Lee, N., Welsh, M., Culler, D.: TOSSIM: accurate and scalable simulation of entire TinyOS applications. In: Proc. Int. Conf. on Embedded Networked Sensor Systems (SENSYS), pp. 126–137 (2003)

[19] Lorincz, K., Welsh, M.: MoteTrack: A robust, decentralized approach to RF-based location tracking. Personal and Ubiquitous Computing 11(6), 489–503 (2007)

[20] Lymberopoulos, D., Lindsey, Q., Savvides, A.: An empirical characterization of radio signal strength variability in 3-D IEEE 802.15.4 networks using monopole antennas. In: Römer, K., Karl, H., Mattern, F. (eds.) EWSN 2006. LNCS, vol. 3868, pp. 326–341. Springer, Heidelberg (2006)

[21] Lynch, C., Reilly, F.O.: Processor choice for wireless sensor networks. In: Proc. ACM Workshop on Real-World Wireless Sensor Networks (REALWSN), pp. 52–68 (2005)

[22] Maróti, M., Völgyesi, P., Dóra, S., Kusý, B., Nádas, A., Lédeczi, Á., Balogh, G., Molnár, K.: Radio interferometric geolocation. In: Proc. Int. Conf. on Embedded Networked Sensor Systems (SENSYS), pp. 1–12 (2005)

[23] Moore, D., Leonard, J.J., Rus, D., Teller, S.J.: Robust distributed network localization with noisy range measurements. In: Proc. Int. Conf. on Embedded Networked Sensor Systems (SENSYS), pp. 50–61 (2004)

[24] Nagpal, R., Shrobe, H.E., Bachrach, J.: Organizing a global coordinate system from local information on an ad hoc sensor network. In: Zhao, F., Guibas, L.J. (eds.) IPSN 2003. LNCS, vol. 2634, pp. 333–348. Springer, Heidelberg (2003)

[25] Nasipuri, A., Li, K.: A directionality based location discovery scheme for wireless sensor networks. In: Proc. ACM Int. Workshop on Wireless Sensor Networks and Applications (WSNA), pp. 105–111 (2002)

[26] Niculescu, D., Badrinath, B.R.: Ad hoc positioning system (APS) using AOA. In: Proc. IEEE Conf. on Computer Communications (INFOCOM), pp. 1734–1743 (2003)

[27] Niculescu, D., Nath, B.: DV based positioning in ad hoc networks. Telecommunication Systems 22, 267–280 (2003)

[28] Ochi, H., Tagashira, S., Fujita, S.: A localization scheme for sensor networks based on wireless communication with anchor groups. In: Proc. Int. Conf. on Parallel and Distributed Systems (ICPADS), pp. 299–305 (2005)

[29] Prasithsangaree, P., Krishnamurthi, P., Chrysanthis, P.K.: On indoor position location with wireless LANs. In: Proc. IEEE Int. Symposium on Personal, Indoor, and Mobile Radio Communications (PIMRC), pp. 720–724 (2002)

[30] Ramadurai, V., Sichitiu, M.L.: Localization in wireless sensor networks: A probabilistic approach. In: Proc. Int. Conf. on Wireless Networks (ICWN), pp. 275–281 (2003)

[31] Rappaport, T.S., Rappaport, T.: Wireless Communications: Principles and Practice, 2nd edn. Prentice-Hall, Englewood Cliffs (2001)

[32] Roos, T., Myllymaki, P., Tirri, H., Misikangas, P., Sievanen, J.: A probabilistic approach to WLAN user location estimation. International Journal of Wireless Information Networks 9(3), 155–166 (2002)

[33] Savvides, A., Han, C.-C., Strivastava, M.B.: Dynamic fine-grained localization in Ad-Hoc networks of sensors. In: Proc. of the Int. Conf. on Mobile Computing and Networking (MOBICOM), pp. 166–179 (2001)

[34] Savvides, A., Park, H., Srivastava, M.B.: The bits and flops of the n-hop multilateration primitive for node localization problems. In: Proc. ACM Int. Workshop on Wireless Sensor Networks and Applications (WSNA), pp. 112–121 (2002)

[35] Shen, X., Wang, Z., Jiang, P., Lin, R., Sun, Y.: Connectivity and RSSI based localization scheme for wireless sensor networks. In: Huang, D.-S., Zhang, X.-P., Huang, G.-B. (eds.) ICIC 2005. LNCS, vol. 3645, pp. 578–587. Springer, Heidelberg (2005)

[36] Sichitiu, M., Ramadurai, V.: Localization of wireless sensor networks with a mobile beacon. In: Proc. IEEE Conf. on Mobile Ad-hoc and Sensor Systems (MASS), pp. 177–183 (2004)

[37] Snyman, J.A.: Practical Mathematical Optimization: An Introduction to Basic Optimization Theory and Classical and New Gradient-Based Algorithms. Springer, Heidelberg (2005)

[38] Stoleru, R., Vicaire, P., He, T., Stankovic, J.A.: StarDust: a flexible architecture for passive localization in wireless sensor networks. In: Proc. Int. Conf. on Embedded Networked Sensor Systems (SENSYS), pp. 57–70 (2006)

[39] Texas Instruments. MSP430C13x1 Datasheet (Revised September 04, 2004)

[40] Wackerly, D., Mendenhall, W., Scheaffer, R.: Mathematical Statistics with Applications, 6th edn. Duxbury Advanced Series (2002)

[41] Whitehouse, K., Karlof, C., Culler, D.: A practical evaluation of radio signal strength for ranging-based localization. In: ACM Mobile Computing and Communications Review, pp. 41–52 (2007)

[42] Xiang, Z., Song, S., Chen, J., Wang, H., Huang, J., Gao, X.: A wireless LAN-based indoor positioning technology. IBM J. Res. Dev. 48(5/6), 617–626 (2004)

[43] Yedavalli, K., Krishnamachari, B., Ravula, S., Srinivasan, B.: Ecolocation: a sequence based technique for RF localization in wireless sensor networks. In: Proc. Int. Conf. on Information Processing in Sensor Networks (IPSN), p. 38 (2005)

[44] Yu, Y., Govindan, R., Estrin, D.: Geographical and energy aware routing: A recursive data dissemination protocol for wireless sensor networks. Technical Report UCLA/CSD-TR-01-0023, UCLA Computer Science Department (2001)

Random Fault Attack against Shrinking Generator*

Marcin Gomułkiewicz[1], Mirosław Kutyłowski[1], and Paweł Wlaź[2]

[1] Wrocław University of Technology
[2] Lublin University of Technology

Abstract. We concern security of *shrinking generator* against fault attacks. While this pseudorandom bitstream generator is cryptographically strong and well suited for hardware implementations, especially for cheap artefacts, we show that using it for the devices that are not fault resistant is risky. That is, even if a device concerned is tamper-proof, generating random faults and analyzing the results may reveal secret keys stored inside the device.

For the attack we flip a random bit and observe propagation of errors. The attack uses peculiar properties of the shrinking generator and presents a new kind of threats for designs based on combining weaker generators. In particular, it indicates that potentially all designs based on combining LFSR generators might be practically weak due to slow propagation of errors in a single LFSR.

1 Introduction

Small sensing devices and other tiny artefacts are crucial for future pervasive systems. While in certain situations they transmit sensitive information, it is necessary to protect their integrity and confidentiality in a reliable way. It is also necessary to authenticate the source of information and secure against attacks that are specific to ad hoc unsupervised systems (such as the replay attack - transmitting once more the old encrypted packets). Theoretically, there are many cryptographic schemes that provide solutions for these problems. However, they are typically designed for different scenarios: broadband communication, reasonable computing power, secure computing units, . . . and are useless for tiny artefacts.

Since certain devices in pervasive systems should be as simple as possible in order to reduce their price, energy consumption and the number of possible faults, it would be desirable to implement basic cryptographic primitives with a simple dedicated hardware. Indeed, these devices may be too weak to operate with a standard processor. *Simple* means here ease of implementation concerning such issues as layout area and fault tolerance. This excludes asymmetric methods, at least according to the current state of technology. So in practice it excludes majority of cryptographic protocols. However, the situation is even worse: solutions based on such primitives as currently used hash functions like SHA-1 turn out to be of little interest due to the cost of hardware implementation. The same applies to standard symmetric encryption methods, regardless of

* This work was supported by Polish Ministry of Science and Education, grant 3T11C 011 26. It was finalized within a project partially supported by EU within the 7th Framework Programme under contract 215270 (FRONTS). An extended abstract of this paper appeared as Dagstuhl report urn:nbn:de:0030-drops-6117.

S. Fekete (Ed.): ALGOSENSORS 2008, LNCS 5389, pp. 87–99, 2008.
© Springer-Verlag Berlin Heidelberg 2008

their careful design and focusing on ease of hardware implementation. In such a situation the only remaining choice is to design security mechanisms based on the simplest stream ciphers. This in turn requires constructing strong pseudorandom bit generators that would be both easy to implement by simple circuits and resistant to various attacks.

Prominent examples of this approach are implementations of algorithms like A5 (for GSM) and Bluetooth (for computer devices). Both solutions fail– the underlying algorithms turn out to be relatively weak even without fault analysis.

Unfortunately, the sensors and other tiny devices cannot be physically protected – they can be captured by an adversary and inspected in a cryptanalytic way. This may be dangerous for a whole system, since the devices often share secret keys. So at least we should make them resistant to simple attacks that can be carried out without sophisticated equipment. The most important issue is proliferation of such attack possibilities. It is critically dangerous, if an attack is possible with widely available and cheap technology.

Shrinking Generator. Recall that an LFSR (Linear Feedback Shift Register) with output $d_1, d_2, d_3, \ldots$ is characterized by the equality

$$d_h = \sum_{i=1}^m t_i \cdot d_{h-m+i} \bmod 2$$

that holds for each $h > m$. The number m is called *length* of the LFSR, the sequence of bits $t_1, \ldots, t_m$ is called the *tap sequence*, $d_1, \ldots, d_m$ is its secret seed.

With all good statistic properties that can be achieved by an LFSR, it is useless in cryptographic sense - breaking it is just solving a set of linear equations. Shrinking generator invented by Coppersmith *et al.* [3] is one of the main designs in the area of pseudorandom bitstream generators. Its advantages are simplicity, high efficiency and relatively high resistance to conventional cryptanalytic attacks. The shrinking generator attempts to create a cryptographically strong pseudorandom bitstream generator out of weak components, usually LFSR's.

Many other solutions of this kind were proved to be weak. [14,15]. The shrinking generator (and its variant self-shrinking generator) successfully faces the trial of time: the best known attacks against it are exponential in the LFSR's length [4,6,7,10,11,12,13], or based on the assumption that the feedback is known [5].

Amazingly, the construction of the shrinking generator is very simple. It consists of two bitstream generators (most frequently LFSRs) we shall call them the input (or base) generator A and the control generator C; their output is denoted by $a_1, a_2, a_3, \ldots$ and $c_1, c_2, c_3, \ldots$, respectively. The output $Z = z_1, z_2, z_3, \ldots$ is composed of those and only those of a_i for which $c_i = 1$. Formally: $z_t = a_i$ for i so that:

$$t = \sum_{j=1}^i c_j, \quad c_i = 1. \tag{1}$$

Fault Attacks. In practice, the attacks can use all available means. The idea [2] of fault attacks is to induce an error in a computation and compare a faulty output with the correct one. In this way, we are sometimes able to derive information on the secret keys contained in the cryptographic device in a non-invasive way. The simplest way to induce errors is to use high level electromagnetic radiation (for instance by putting a piece of uranium on the device). A particle intruding the chip changes the contents of a register to a low energy state (so typically it changes a one to a zero). The change concerns one bit or a group of neighboring bits (depending on the technology used and size of the registers).

Previous Results. The paper [9] shows two fault attacks against the shrinking generator. The first one is based on the assumption that the clocks of the internal generators can be desynchronized (a similar assumption was made in [8]), the second one is essentially based on the possibility of replacing the control register by a source of independent, random bits, while keeping intact the contents of the input register. The first attack gives quite powerful results (with high probability only a couple of possible control sequences, including the correct one), but the assumptions made require the cooperation of the device's manufacturer (and/or very careless design of the device). The second attack, while quite feasible from a technical point of view, gives only moderate results – it yields a sequence that coincides with the input sequence of the shrinking generator on a large fraction of positions. However, it is not known on which positions both sequences coincide. So, this attack should be regarded as an initial phase of a possible full attack revealing the internal state of the shrinking generator.

New Results. We propose attacks based on analysis of propagation of errors caused by single bit flips. These attacks are technically more feasible and hard to prevent, unless special error propagation countermeasures are developed (which is not the case yet). Our goal is to show certain novel approaches to fault analysis and in this way give a chance to respond with new algorithms tailored for sensor devices that would be sufficiently immune against fault attacks and cryptographically strong at the same time.

2   Attack by Faults in the Input Register

In this section we consider a shrinking generator, for which the input register is an LFSR with a known feedback, while the control register is an arbitrarily chosen bitstream generator. Our goal is to find the contents of the control register (which typically depends directly on the shared key).

Assumptions. Let A denote the input register. We assume that A is an LFSR of length m with a known feedback function f. Let its initial contents be denoted by $a_1, a_2, \ldots, a_m$ (a_1 being on the output position). Let Z denote the output of the shrinking generator. We assume that we can restart the shrinking generator and flip exactly one of bits $a_i, i \leq m$ and get a new output sequence Z'. In this case the control register starts with exactly the same contents as before. We assume that the error introduced is transient – after restarting the device we get the original contents of the input register.

By f we denote the known feedback function of the input sequence, and let F describe the output of A, that is $F \colon \{0,1\}^m \to \{0,1\}^\infty$ and

$$F(a_1, a_2, \ldots, a_m) = (a_1, a_2 \ldots) \quad \text{where} \quad a_{p+m} = f(a_p, \ldots, a_{p+m-1}), \text{ for } p \geq 1.$$

Since we assume f to be a linear function, also F is linear, that is

$$F(\mathbf{a}) + F(\mathbf{b}) = F(\mathbf{a} + \mathbf{b}), \tag{2}$$

where $+$ denotes sum modulo 2. Sometimes f is described by a so called *tap sequence* $t_0, t_1, \ldots, t_{m-1}$ such that $t_j \in \{0, 1\}$ and $t_0 = 1$:

$$f(a_p, a_{p+1}, \ldots, a_{p+m-1}) = \sum_{j=0}^{m-1} t_j \cdot a_{p+j} \, .$$

Sketch of the Attack. Suppose that we have flipped a_k, for some $i \leq m$, and that the resulting output of the input register is $a_1', a_2', \ldots$. According to (2) it means that

$$
\begin{aligned}
(a_1', a_2', \ldots) &= F(a_1', a_2', \ldots, a_m') = \\
&= F(\mathbf{e}_k + (a_1, a_2, \ldots, a_m)) = \qquad (3) \\
&= F(\mathbf{e}_k) + (a_1, a_2 \ldots)
\end{aligned}
$$

where

$$\mathbf{e}_k = (\underbrace{0, 0, \ldots, 0}_{k-1}, 1, \underbrace{0, 0, \ldots, 0}_{m-k}).$$

We assume that we know exactly the value of k. If this is not the case, then complexity of the attack will increase by a small factor bounded by m. (Moreover, it is possible to estimate k for sparse tap sequences using methods described in the next section.)

It follows from (3) that by adding $(a_1, a_2, \ldots)$ and $(a_1', a_2', \ldots)$ we obtain $F(\mathbf{e}_k)$. So $F(\mathbf{e}_k)$ gives us positions where the sequences $(a_1, a_2, \ldots)$ and $(a_1', a_2', \ldots)$ differ. However, an attacker can see only $(z_1, z_2, \ldots)$ and $(z_1', z_2', \ldots)$. Now we will try to reconstruct n consecutive bits of the bitstream $(c_1, c_2, \ldots)$ generated by the control register, where n may be chosen arbitrarily. If we knew the feedback of the control register, then of course we could choose n to be equal its length and try to obtain the $c_q, c_q + 1, \ldots, c_{q+n-1}$ for some known $q \geq 1$.

For a moment we fix some p and q. We consider output sequences of the shrinking generator from positions p and will assume that these both sequences correspond to the input sequence and the control sequence from position q. Then $p = q/2 + \Delta$, where Δ is small value compared to q and can be estimated with Chernoff bounds as for the number of successes in q Bernoulli trials.

In a standard way we construct a labeled tree describing the possible values of $c_q, c_{q+1}, \ldots$ – if the edges of a path from the root to a node s have labels $b_1, \ldots, b_u$, then s corresponds to the situation that $c_1 = b_1, \ldots, c_u = b_u$ (see Fig. 1).

Suppose that we have guessed all the values $c_q, c_{q+1}, \ldots, c_{q+i-1}$. We want to guess the value of c_{q+i}. One of the choices is of course 0, but when can it be 1? If w denotes the number of ones in $c_q, c_{q+1}, \ldots, c_{q+i-1}$, then we can exclude the value $c_{q+i} = 1$ if

$$F(\mathbf{e}_k)|_{q+i} \neq (z_{p+w} + z_{p+w}' \bmod 2) \qquad (4)$$

This moment is quite important, since if the probability of (4) is fairly large, then we will be able to prune a substantial number of subtrees of the tree describing all possible values of the control sequence.

First assume that the guess of the control sequence up to position $q+i-1$ was correct so far in the sense that the number of ones up to c_{q+i-1} is correct (but not necessarily their distribution). Consider a node s of the tree corresponding to this choice. Since we cannot exclude the possibility that $c_{q+i} = 0$, node s has a son to which leads an edge with label 0. On the other hand, if $c_{q+i} = 1$ and this is the right choice, then the inequality (4) does not hold and s has the second son. If this is not the right choice, then (4) does not hold if and only if $F(\mathbf{e}_k)|_{q+i} = (z_{p+w} + z'_{p+w} \bmod 2)$, while in fact we have

$$(z_{p+w} + z'_{p+w} \bmod 2) = F(\mathbf{e}_k)|_{q+i+h} \ ,$$

where $h \geq 0$ is the smallest index such that $c_{q+i+h} = 1$. So, node s has the second son if and only if

$$F(\mathbf{e}_k)|_{q+i} = F(\mathbf{e}_k)|_{q+i+h} \ . \tag{5}$$

At this point we may assume that LFSR used as the input generator has good statistical properties and from some position $F(\mathbf{e}_k)$ passes standard statistical tests. So in average conditions like (5) are satisfied for 50% of cases, provided that q is large enough.

Now consider the remaining cases. As before, we cannot prune the son of s corresponding to the guess that $c_{q+i} = 0$. On the other hand, for $c_{q+i} = 1$ we check the condition (4). It concerns the value $(z_{p+w} + z'_{p+w} \bmod 2)$ which equals $F(\mathbf{e}_k)|_{q+j}$ for some j that is different from i. So, the condition (4) is satisfied in this case iff

$$F(\mathbf{e}_k)|_{q+i} = F(\mathbf{e}_k)|_{q+j} \ . \tag{6}$$

Due to the statistic properties of the input register, this occurs in about 50% of cases.

It follows from our considerations that the expected number of sons of a node of the search tree is slightly higher than $\frac{3}{2}$. Since the height of the tree is n, we expect about

$$\left(\tfrac{3}{2}\right)^n \approx 2^{0.585n} \tag{7}$$

leaves in the tree. So for $n = 64$ we get a tree with about $2^{37.5}$ leaves and gain about $0.415 \cdot n \approx 26.5$ bits. For $n = 94$ we get 2^{55} leaves and gain about $0.415 \cdot n \approx 39$ bits. Of course, this gain is smaller, if we cannot estimate well p and q.

Example. We explain the above ideas with the following toy example. Let

$$f(x_1, x_2, \ldots, x_6) = x_1 + x_2 \bmod 2 \tag{8}$$

(so $a_p = a_{p-6} + a_{p-5}$, for $p > 6$). Assume that the control sequence is generated by some other LFSR of length 5. Let us have, in our example, two output sequences:

$$\begin{aligned}(z_1, z_2, \ldots) &= 111011110001 \ldots \quad \text{and} \\ (z'_1, z'_2, \ldots) &= 111001101111 \ldots\end{aligned} \tag{9}$$

Assume that the 3rd position in input register (counting from the output position) is flipped, so by (8), $F(\mathbf{e}_k) = 0010000110001010011110100011001001011011101100$ $1101010111110000010000\ldots$ We take $q = 9$, as $F(\mathbf{e}_k)$ is already "random" from position 9 and the number of ones approximately equals the number of zeroes. In the first 8 elements of the control sequence we expect about 4 ones, so we skip the first 4 elements of output sequence. This might be false, but if the attack fails, then we will try other less probable possibilities. First we compute

$$(z_5, z_6, \ldots) + (z'_5, z'_6, \ldots) = 10011110 \ldots \tag{10}$$

Note that
$$(F(\mathbf{e}_k)|_9, F(\mathbf{e}_k)|_{10}\ldots) = 10001010011110100\ldots$$

Now consider the tree of guesses for the control sequence from position 9. As always, c_9 can be equal to 0, but $c_9 = 1$ is also possible since $F(\mathbf{e}_k)|_9 = (z_5 + z_5' \bmod 2)$. If $c_9 = 0$, then $c_{10} \neq 1$ as $F(\mathbf{e}_k)|_{10} \neq (z_5 + z_5' \bmod 2)$. We continue in that manner until we obtain a tree of height 5 (five is the length of control register) – where every leaf gives us values of $c_9, c_{10}, \ldots, c_{13}$ (see Fig. 1).

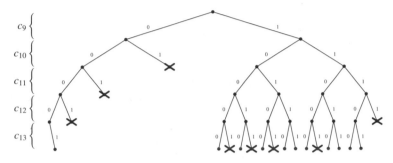

Fig. 1. A tree of guesses for the control sequence in the example

The tree constructed has only 11 leaves (note that the leaf corresponding to $00\ldots0$ is excluded, as this is the content of the register which never appears in calculations). Each leaf corresponds to a full contents of the control register. To proceed we assume to know the feedback of the control LFSR, for instance
$$c_i = c_{i-5} + c_{i-3} \bmod 2$$

Now we can emulate the system (taking $\mathbf{e}_k$ as the starting state for the input register) and compare the results with $z + z' \bmod 2$. Take the first leaf corresponding to $c_9 c_{10} .. c_{13} = 00001$. Then $c_9 \ldots c_{18} = 0000100101$ and, consequently, $z_7 = a_{18}$. On the other hand, $F(\mathbf{e}_k)|_{18} = 1$, what implies that $a_{18} \neq a_{18}'$. However, in our example $z_7 = z_7'$, thus we have to exclude the leaf corresponding to 00001.

In a quite similar way we exclude all but one leaf, namely 11101, so this is the only candidate left for the true contents of $c_9 c_{10} c_{11} c_{12} c_{13}$. We clock the control register back and obtain $c_1 c_2 c_3 c_4 c_5 = 11000$. Now we can check that among the first 8 control bits there are exactly 4 ones, as assumed at the beginning. Now we can write equations for a_i, and since we can express all a_i as linear expressions of $\{a_1, a_2, a_3, a_4, a_5, a_6\}$, we need 6 independent equations. For instance,
$$a_1 = z_1, \quad a_2 = z_2, \quad a_6 = z_3, \quad a_3 + a_4 = z_5, \quad a_5 + a_6 = z_6, \quad a_1 + a_3 = z_8$$

which leads to a solution $a_1 a_2 a_3 a_4 a_5 a_6 = 110101$.

Experimental Data. Since formula (7) has been obtained by some simplifications we have implemented the attack and checked performance of the attack in practice.

In the experiments we choose at random 32-bit LFSRs with 4 taps for the control registers and 32-bit LFSRs with 6 taps for input registers (all of them with the period

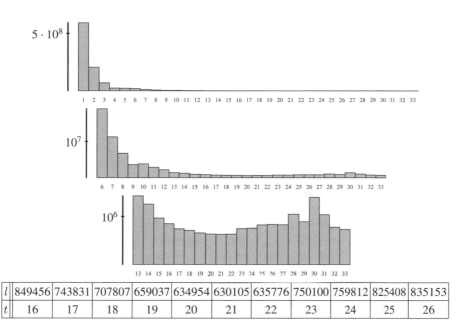

l	849456	743831	707807	659037	634954	630105	635776	750100	759812	825408	835153
t	16	17	18	19	20	21	22	23	24	25	26

Fig. 2. The number l of leaves in a search tree related to the number t of ones in a segment of $F(\mathbf{e}_k)$ of size 41

$2^{32} - 1$). We considered about 800 such pairs and for each pair we took as a number q anything from 20 to 200. As starting contents of the control register we gave some random value and run the simulator. For each value of q we considered all p between $q/2 - 10$ and $q/2 + 10$ and calculated the average number of leaves in each case.

We experimentally investigate what is the *sufficiently large* q. We look for a simple rule to make a decision. For this purpose we consider the number t of ones within the segment $[q - 20, q + 20]$ in the sum of the input sequence and the faulty input sequence. For each t we have computed the average size l of the search tree. As one might expect t equal to about half of the width of the segment leads to optimum number of leaves of tree – see Fig. 2. On this figure at the bottom of each bar we have marked the number t, the height of the bar reflects the average number of leaves. Due to high differences in the values of l, we have split the picture into 3 parts, each in a different scale. In the lowest part of Fig. 2 we present some exact values. We see that applying the rule that the number of ones in a segment of $F(\mathbf{e}_k)$ should be about the half of the length of the segment turns to be a good choice: the average number of leaves for $t \in \{20, 22\}$ are approximately $(1.52)^{32}$, which is close to (7).

3 Attack through Faults in Control Register

Now we outline another attack against the shrinking generator. This time, unlike in [9], we shall assume that the control register is an LFSR with a known feedback, while the input generator may be arbitrarily chosen random bit generator. Our goal is to reconstruct the contents of the control register. Of course, once this is achieved and the input register is also an LFSR generator, one can immediately reconstruct the contents of the input register.

Notations and Assumptions. We shall adopt the notations from Section 2. On top of that we shall assume that one can cause exactly one random bit-flip within the control register and rerun the generator. However, the position of the bit flipped is random. This corresponds to physical reality of radiation-induced bit-flips – neutrons are able to introduce transient and permanent faults. That is, if the register has length n, and its cells contain $c_1, c_2, \ldots, c_n$ (c_1 in the output position), then it is possible to get the output sequence Z' corresponding to the $c'_1, c'_2, \ldots, c'_n$, where all but one c'_i are equal to c_i.

Attack Idea. For the sake of simplicity we present the attack for a specific case, where it is particularly evident how it works and the number of possible details and attack options is relatively small. We shall consider one of the tap sequences listed in [1], namely LFSR register of length 94 with tap sequence 94, 73 (i.e. we have $c_i = c_{i-94} + c_{i-73}$, for all $i > 94$). Let us call it L.

Let us use the following convention: let c_j denote the bit which is shifted along L, so that it would become the jth output of L, if L were used alone as pseudorandom generator. The first observation is that if we flip c_s at the moment when c_s is before position 74 in L, then c_{s+73}, c_{s+94} will be flipped as well due to changes induced directly by c_s through the tap sequence, and $c_{s+73+73}, c_{s+73+94}$ due to changes induced directly by c_{s+73} through the tap sequence, and $c_{s+94+73}, c_{s+94+94}$ due to changes induced directly by c_{s+94} through the tap sequence, and

$$c_{s+146+73}, c_{s+146+94}, \quad c_{s+188+73}, c_{s+188+94}, \quad c_{s+219+73}, c_{s+219+94}$$

and so on. Note that finally changes occur at

$$c_s, c_{s+73}, c_{s+94}, c_{s+146}, c_{s+188}, c_{s+219}, c_{s+240}, c_{s+261}, c_{s+282}, \ldots$$

(notice that two flips occurring at c_{s+167} cancel themselves). We see that (at least at the beginning) the distances between the changed bits are quite large.

Let $a_1, a_2, \ldots$ denote the output that would be generated by the input generator. The second observation is that flipping a control bit c_j can have two different effects:

1. if a zero is replaced by one – then an additional element a_j from the input generator is inserted into the output sequence, or
2. a one is replaced by a zero – in this case a_j is removed from the output sequence.

It turns out that it is possible to detect which case has occurred, and so in this way find out what was the value of the bit flipped. Consider first the effect of flipping c_s. Since the bits up to c_s are unchanged, until the position corresponding to s, the output of the shrinking generator remains the same. Let t be this position. Then either a bit is removed from position t or a new bit is inserted between positions $t - 1$ and t. The output bit from position t occurs in a block of bits of the same value (in particular, it may consist of this single bit). Without loss of generality assume that this is a block

correct output:

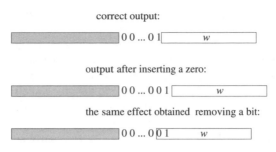

Fig. 3. Possible interpretation of a change in the output sequence, when c_s is changed from 0 to 1

of zeroes. If such a block gets longer, then two reasons are possible (see Fig. 3). One is inserting an additional zero. The second one is that a one separating two blocks of zeroes is removed. In some distance no other control bit is affected by the change of c_s, so a certain sequence w remains unchanged after the mentioned block of zeroes is in the output of the shrinking generator. One can easily see that we cannot resolve which case has occurred, if w is a prefix of $10w$. This occurs only if $w = 101010\ldots$. If $|w|$ is large (say $|w| \geq 20$), the case of such a pattern occurs quite infrequently. Small values of $|w|$, on the other hand, also happen with very low probability. Indeed, we can assume that the output of the input generator has good stochastic properties. Then the average value of $|w|$ is about $72/2 = 36$ and $|w| < 20$ with probability less than $4 \cdot 10^{-5}$. So the case when we cannot distinguish between both cases occurs with a very low frequency.

If the block of zeroes concerned gets shorter, then again there are two cases. Obviously, it can be due to removing a bit from the output of the shrinking generator. The other case is that a one is inserted in front of the last zero in the block (see Fig. 4). In this case w must be a prefix of $01w$, hence $w = 0101\ldots$.

We see that with a high probability we may determine which change has occurred via flipping a bit, and so what was the value of the bit flipped. What we do not know is the position of the flipped bit in the control register. We even cannot say exactly which bit of the output sequence has been removed or inserted, respectively. We can only point to a block of the same bit value where the change has occurred (of course, we can locate the changes in cases when the block of length one is canceled or a new block is

correct output:

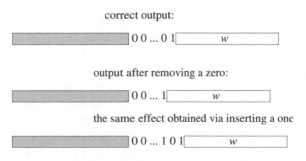

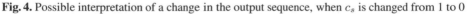

Fig. 4. Possible interpretation of a change in the output sequence, when c_s is changed from 1 to 0

introduced). Hence, so far we do not get much useful information for reconstructing the contents of the control register.

Reconstructing Induced Changes. One of the crucial observations is that in the scenario considered above flipping the bit c_s causes flipping the bits c_{s+73}, c_{s+94}, c_{s+146}, c_{s+188}, ... as well. Due to large distances between the bits flipped with high probability we can recognize their values as described in the previous subsection (at least at the beginning, since the bits flipped become eventually quite dense). Note that we know exactly the distances between these bits in the output of the control register. So, each time we flip a bit, we get a pattern of bits c_s, c_{s+73}, c_{s+94}, c_{s+146}, c_{s+188}, c_{s+219}, c_{s+240}, c_{s+261}.

In our example, we can reconstruct a few more bits in this way. In fact, notice that $c_s + c_{s+21} = c_{s+94}$, $c_{s+52} + c_{s+73} = c_{s+146}$, $c_{s+94} + c_{s+115} = c_{s+188}$, $c_{s+125} + c_{s+146} = c_{s+219}$. Hence we can derive easily the values of $c_{s+21}, c_{s+52}, c_{s+115}, c_{s+125}$. Then in turn we consider equality $c_{s+31} + c_{s+52} = c_{s+125}$, and derive c_{s+31}. Similarly, $c_{s+22} + c_{s+43} = c_{s+115}$, hence we may derive c_{s+43}.

So finally we get a pattern consisting of the values

$$c_s, c_{s+21}, c_{s+31}, c_{s+43}, c_{s+52}, c_{s+73}, c_{s+94}, c_{s+115}, c_{s+125}, c_{s+146}, \cdots$$

For our example, we shall use the values of c_s, c_{s+21}, c_{s+31}, c_{s+43}, c_{s+52}, c_{s+73}.

The main disadvantage concerning the indexes of bits recovered according to the procedure described in the previous subsection is that they do not occur in the same block of length 94. Nevertheless, as we have shown, we can "shift" information by considering the tap connections. This holds also in the case when tap sequences consist of more than two connections. The only disadvantage in this case is that instead of values of single bits we get values of sums of bits. Treating these as a system of linear equations for $c_s, \ldots, c_{s+93}$ we will in fact gain as many bits as is the rank of the system (by guessing some values and calculating all other occurring in the sums).

Ordering Different Patterns. Assume that we have recovered two patterns, say

$$c_s, c_{s+73}, c_{s+94}, c_{s+146}, c_{s+188}, \cdots \text{ and } c_z, c_{z+73}, c_{z+94}, c_{z+146}, c_{z+188}, \cdots,$$

but we do not know the values of s and z. We shall see that almost always we may recognize which of the values s, z is smaller.

The changes of c_s and c_z cause changes in the output of the shrinking generator. As above, the place of a change can be located within a block of the same symbols. Obviously, if the change related to c_z occurs in a block that comes later than the block, where a change is caused by flipping c_s, then obviously $s < z$.

If we cannot determine that $s < z$ or $z < s$, since the first changes occur in the same block of the output of the shrinking generator, we still have chance to compare s and z by considering the changes in the output of the shrinking generator induced by the changes that occur at c_{s+73} and c_{z+73}. If this fails, we have still a chance to detect the order of s and z at c_{s+94} and c_{z+94}, c_{s+146} and c_{z+146}, c_{s+188} and c_{z+188}, c_{s+219} and c_{z+219}, c_{s+240} and c_{z+240}, c_{s+261} and c_{z+261}, c_{s+282} and c_{z+282}. Observe that if the bits generated by the input register on positions $s + j$ and $z + j$ are different, we shall detect the order of s and z. So if we treat the output of the input generator as a random string, then with probability only 2^{-9} the opposite case occurs for each of the 9 pairs listed. In fact, probability of detection is higher: even if the bits generated by the input generator on positions $s + j$ and $z + j$ are equal, it may happen that they do not belong to the same block of identical bits in the output generated by the correct computation. Probability of such an event grows with the distance between s and z.

Let us observe that not only we can detect that, for instance $s < z$, but also get further estimation, like $z < s+21$. For this purpose we look at the changes induced by changing c_{z+73} and c_{s+94}. As before, if these changes do not occur in the same block of the same symbols in the output of the shrinking generator, we can derive this information. Similarly, in order to get analogous results we may consider the changes induced by c_{z+188} and c_{s+219}, c_{z+219} and c_{s+240}, c_{z+240} and c_{s+261}, c_{z+261} and c_{s+282}. Hence, with a fairly good probability we may detect patterns generated by flipping c_z where $z \in (s, s + 240)$.

Additional Data – Number of Ones. Analyzing errors induced for generating patterns discussed in the previous subsection yields one more information. Consider for instance c_{s+73} and c_{s+94}. We can determine the place where the change of c_{s+73} and c_{s+94} yields the changes in the output of the shrinking generator. These places can be determined almost exactly – uncertainty is related to the case when we remove or introduce symbols within a block of the same bits in the output string. The best case is that this block is of length one – then we know the position exactly. Anyway, the average length of such a block is less than 2.

On the other hand, the number of bits in the output of the shrinking generator between the positions corresponding to c_{s+73} and c_{s+94} equals the number of ones in the sequence c_{s+74} through c_{s+93}. This reduces the number of cases for a brute force search. For instance, instead of 2^{20} possible settings we have to consider only $\binom{20}{w}$ cases, where w is the number of ones. For $w = 10$ we get the maximum number of cases $\approx 2^{17.5}$.

Sketch of the Attack. Let us now summarize the steps performed during the attack:

1. We generate 1 bit failures in the control generator in about the same time, each time restarting the shrinking generator in the same state.
2. We choose a group of patterns such that:
   - if the pattern with the earliest error corresponds to flipping c_s, then for each other pattern of the group the bit flipped is not later than c_{s+21},
   - we can determine the time ordering between each pair of patterns in the group,
   - c_s corresponds to a block of length 1 (in fact, this only simplifies the description),
   - the group chosen has the highest cardinality among the groups having the above properties.
3. If $s + s_1 < s + s_2 < \ldots < s + s_d$ denote the indexes of the bits flipped for the patterns in the group, we consider separately each configuration $0 < s_1 < \ldots < s_d < 21$.
4. Given a configuration $0 < s_1 < \ldots < s_d < 21$ we set all remaining bits in positions s through $s + 93$ without violating the conditions about the number of ones in each of the intervals.

5. In order to check a solution we generate the outputs of the control register for a few hundred of bits for each of the patterns. Then we check if they are consistent with the outputs of the shrinking generator. That is, if for s_i and s_j we have $c_l = 1$, then we check if the output bits of the shrinking generator are the same on positions corresponding to c_l for both erroneous outputs. Note that the position in the output string corresponding to c_l is the position of c_s plus the offset given by the number of ones in the sequence $c_{s+1}, \ldots, c_l$.

Let us comment on the computational complexity of this attack. Assuming that we get 8 patterns of the properties described we have to consider $\binom{20}{7} \approx 2^{16}$ cases in step 3. However, in this way we set $8 \times 6 = 48$ bits. Setting the remaining bits requires considering $2^{93-48} = 2^{45}$ cases. In fact, this is overestimated, since we can bound the number of cases based on the information of the number of ones in certain intervals – the search can be performed using the standard branch and bound technique. Together we have a rough estimate of 2^{61} cases to be considered. So the gain of the attack in this case is more than 33 bits, even if the brute force attack could consider the control register alone.

Final Remarks

Even if in our presentation we assume that the input and control sequences are generated by LFSRs, in fact we require a slightly weaker assumption. Namely, if we flip a single bit of the internal state of a generator, then we can say which of the bits of the output of the generator become flipped as well.

It follows from the discussion in the paper that in most cases one can easily detect that the number of bits flipped is higher than 1 and, for instance, skip such data. On the other hand, the attack can be adapted to the case when not a single bit but a group of a few consecutive bits is flipped at once. This would correspond to realistic fault conditions for the highest integration scale. We also have to remark that even if the attacks presented are focused on some specific details, the case study shows that there might be many ways to fine tune the attack.

References

1. Alfke, P.: Efficient Shift Registers, LFSR Counters, and Long Pseudo-Random Sequence Generators. Application Note, XAPP 052 July 7 (1996) (Version 1.1), http://www.xilinx.com/bvdocs/appnotes/xapp052.pdf
2. Boneh, D., DeMillo, R.A., Lipton, R.J.: On the Importance of Checking Cryptographic Protocols for Faults. In: Fumy, W. (ed.) EUROCRYPT 1997. LNCS, vol. 1233, pp. 37–51. Springer, Heidelberg (1997)
3. Coppersmith, D., Krawczyk, H., Mansour, Y.: The Shrinking Generator. In: Stinson, D.R. (ed.) CRYPTO 1993. LNCS, vol. 773, pp. 22–39. Springer, Heidelberg (1994)
4. Dawson, E., Golič, J.D., Simpson, L.: A Probabilistic Correlation Attack on the Shrinking Generator. In: Boyd, C., Dawson, E. (eds.) ACISP 1998. LNCS, vol. 1438, pp. 147–158. Springer, Heidelberg (1998)
5. Ekdahl, P., Johansson, T., Meier, W.: Predicting the Shrinking Generator with Fixed Connections. In: Biham, E. (ed.) EUROCRYPT 2003. LNCS, vol. 2656, pp. 330–344. Springer, Heidelberg (2003)

6. Golić, J.D., O'Connor, L.: Embedding and Probabilistic Correlation Attacks on Clock-Controlled Shift Registers. In: De Santis, A. (ed.) EUROCRYPT 1994. LNCS, vol. 950, pp. 230–243. Springer, Heidelberg (1995)
7. Golić, J.D.: Correlation Analysis of the Shrinking Generator. In: Kilian, J. (ed.) CRYPTO 2001. LNCS, vol. 2139, pp. 440–457. Springer, Heidelberg (2001)
8. Gomułkiewicz, M., Kutyłowski, M., Vierhaus, T.H., Wlaź, P.: Synchronization Fault Cryptanalysis for Breaking A5/1. In: Nikoletseas, S.E. (ed.) WEA 2005. LNCS, vol. 3503, pp. 415–427. Springer, Heidelberg (2005)
9. Gomułkiewicz, M., Kutyłowski, M., Wlaź, P.: Fault Cryptanalysis and the Shrinking Generator. In: Àlvarez, C., Serna, M. (eds.) WEA 2006. LNCS, vol. 4007, pp. 61–72. Springer, Heidelberg (2006)
10. Krause, M.: BDD-based Cryptanalysis of Keystream Generators. In: Knudsen, L.R. (ed.) EUROCRYPT 2002. LNCS, vol. 2332, pp. 222–237. Springer, Heidelberg (2002)
11. Krause, M., Lucks, S., Zenner, E.: Improved Cryptanalysis of the Self-Shrinking Generator. In: Varadharajan, V., Mu, Y. (eds.) ACISP 2001. LNCS, vol. 2119, pp. 21–35. Springer, Heidelberg (2001)
12. Meier, W., Staffelbach, O.: The Self-shrinking Generator. In: De Santis, A. (ed.) EUROCRYPT 1994. LNCS, vol. 950, pp. 205–214. Springer, Heidelberg (1995)
13. Mihaljevic, M.: A Faster Cryptanalysis of the Self-shrinking Generator. In: Pieprzyk, J.P., Seberry, J. (eds.) ACISP 1996. LNCS, vol. 1172, pp. 182–188. Springer, Heidelberg (1996)
14. Rao, T.R.N., Yang, C.-H., Zeng, K.: An Improved Linear Syndrome Algorithm in Cryptanalysis With Applications. In: Menezes, A., Vanstone, S.A. (eds.) CRYPTO 1990. LNCS, vol. 537, pp. 34–47. Springer, Heidelberg (1991)
15. Zenner, E.: On the Efficiency of the Clock Control Guessing Attack. In: Lee, P.J., Lim, C.H. (eds.) ICISC 2002. LNCS, vol. 2587, pp. 200–212. Springer, Heidelberg (2003)

Probabilistic Protocols for Fair Communication in Wireless Sensor Networks*

Ioannis Chatzigiannakis[1,2], Lefteris Kirousis[1,2], and Thodoris Stratiotis[2]

[1] Research Academic Computer Technology Institute, Patras, Greece
[2] Department of Computer Engineering and Informatics, University of Patras, Greece
ichatz@cti.gr, kirousis@ceid.upatras.gr, stratiot@ceid.upatras.gr

Abstract. In this work we present three new distributed, probabilistic data propagation protocols for Wireless Sensor Networks which aim at maximizing the network's operational life and improve its performance. The keystone of these protocols' design is *fairness* which declares that fair portions of network's work load should be assigned to each node, depending on their role in the system. All the three protocols, EFPFR, MPFR and TWIST, emerged from the study of the rigorously analyzed protocol PFR. Its design elements were identified and improvements were suggested and incorporated into the introduced protocols. The experiments conducted show that our proposals manage to improve PFR's performance in terms of *success rate, total amount of energy saved, number of alive sensors* and *standard deviation of the energy left*. Indicatively we note that while PFR's success rate is 69.5%, TWIST is achieving 97.5% and its standard deviation of energy is almost half of that of PFR.

1 Introduction

The subject of this study is the communication protocols for static, homogeneous Wireless Sensor Networks (WSN's), where the energy reserves of the particles are finite. A protocol's goal should be to prolong system's *functional life* and maximize the number of events which get successfully reported. There is a great variety of different protocols for WSN's. A well known data aggregation paradigm is *Directed Diffusion* whose key element is data propagation through multiple paths. *Multi-path* (e.g. [3,7]) protocols contrary to the *single-path* ones, broadcast to more than one nodes in each single hop. Thus they are less affected by node failures and ensure successful message deliveries. Additionally they are easier to implement since they do not involve a route discovery process. On the other hand, they tend to spend more energy than the single-path protocols since they engage more particles to the propagation process than the single-path ones.

Energy Awareness is another important property of communication algorithms applied to sensor networks and many protocols take energy consumption into account for their routing decisions (e.g. [7,9,12]). Some approaches even

---

* Partially supported by the EU within the 7th Framework Programme under contract 215270 (FRONTS).

S. Fekete (Ed.): ALGOSENSORS 2008, LNCS 5389, pp. 100–110, 2008.

propose *sleep/awake* schemes aiming at energy saving. Moreover there are many protocols presented, applying techniques in order to achieve *Balanced Energy Dissipation*. This feature's concept is that, even more important from saving energy, is assuring that the energy is dissipated uniformly by all the particles. Thus the premature depletion of certain keystone nodes, which in turn results to the destruction of the whole network, is avoided and the system operational life is elongated. This goal is pursued by many researchers (e.g. [5,7,12]).

1.1   Protocol Engineering

Our objective is to devise communication protocols for WSN's that allow the underlying network to operate effectively for the longest possible time. These protocols have to be decentralized and adaptive to current conditions. On the other hand they should take measures against the nodes' energy dissipation and their "death". Our efforts were initially focused on the study of the rigorously analyzed *Probabilistic Forwarding Protocol* - PFR [3]. This protocol, presented in more detail in section 4, attracted our attention because of its design characteristics. It is simple, lightweight, probabilistic and totaly distributed. All it needs is local information which is easily provided to the particles, although some researchers [6] suggest the use of global topology, which can be encoded and consequently disseminated throughout the network. On top of that PFR is thoroughly analyzed considering correctness, energy efficiency and robustness.

Then we tested PFR on diverse operation scenarios and we identified its design flaws and its beneficial elements. We modified its components and experimentally assessed their impact. The process yielded interesting ideas which were simulated and evaluated as well. Corrections were suggested and experiments took place to identify the best setting of their parameters. From this procedure protocols EFPFR and MPFR emerged (see sections 5.1 and 5.2 respectively) as descendants of PFR, and protocol TWIST (see section 5.3) evolved, in turn, from MPFR. The keystone of these new protocols' design is *Fairness*. This concept, in the context of WSN's, has been addressed by many researchers in the past (e.g. [10]). In this work the term implies that the protocols must be able to identify each particle's particularities and roles and consequently treat them respectively. For instance important nodes should be relieved from some of their burden, while the seemingly less important ones, should be prompted to participate more in the data propagation process. This could be realized either by prohibiting transmissions from the frequently used, "poor", in terms of energy particles, or by encouraging propagations through paths different from the greedy, shorter ones. The goal is to amplify "solidarity" within the network and keep it functional for more time.

2   The Model and the Problem

In this work we follow the model presented in [3]. It is comprised by N sensor particles and one sink. The particles' positions are chosen uniformly at random.

Sink's resources (energy, transmission radius and processing and storage capabilities) are infinite. Once the sensors are deployed, no new ones are added. The nodes are assumed to be static, identical and their transmission ranges $\mathcal{R}$ and initial energy reserves E_{in}, are finite and fixed. The transmission ranges could be variable as well, but we prefer to deal with a simpler and stricter case. Each data transmission is a broadcast and every particle can communicate with every other being positioned within its range $\mathcal{R}$ assuming the absence of obstacles. A critical event $\mathcal{E}$ occurs with probability $P_{\mathcal{E}}$, called *injection rate*. Additionally we assume that every particle can estimate the direction of a received transmission and the distance from its source, as well as the direction towards the sink. The particles can retrieve this kind of information employing various techniques. One way is incorporating smart antennas into the particles, which may be feasible in the near future (see [4]). Thus within an initialization round, during which the sink is broadcasting a powerful signal of known intensity, the sensors calculate their direction in relation to the sink and their distance from it, measuring signal attenuation. The particles can also use GPS transceivers to estimate their positions or even localization methods (e.g. [11]) to decide on *fictitious virtual coordinates* and thereafter be able to estimate directions and distances.

The Energy Model: As it is proposed in [8], we consider that energy is dissipated when the particles transmit, receive and remain in idle state. The energy needed for a transmission/reception is proportional to the time the transmitter/receiver must operate in order to transmit/receive the entire message. Thus $E_{tx} = T_{tx_{start}} \cdot P_{tx_{start}} + \frac{n_{mes}}{R_t} \cdot (P_{tx_{elec}} + P_{amp})$ is the amount of energy required to transmit a message having length n bits. $T_{tx_{start}}$ is the time required to turn on the transmitter, $P_{tx_{start}}$ is the power needed for this task, R_t is the transmission rate and $P_{tx_{elec}}$ and P_{amp} are respectively the amounts of power needed to operate the transmitter and the amplifier. Similarly the energy needed to receive a message of n_{mes} bits is: $E_{rx} = T_{rx_{start}} \cdot P_{rx_{start}} + \frac{n_{mes}}{R_t} \cdot P_{rx_{elec}}$, where $T_{rx_{start}}$ is the time required to turn on the receiver, $P_{rx_{start}}$ is the power needed to be spent for that purpose and $P_{rx_{elec}}$ is the amount of power required to operate the receiver. Finally when a particle is idle we consider that it spends an amount of energy E_{idle}, equal to that required to receive only one bit. So: $E_{idle} = T_{rx_{start}} \cdot P_{rx_{start}} + \frac{1}{R_t} \cdot P_{rx_{elec}}$. Intuitively, during each time step, a particle switches on its receiver, senses the carrier (receives $O(1)$ bits) and if no worthy transmission is taking place, it switches off the receiver. In some architectures the amount of power the particle's circuit spends during the idle state $(P_{id_{elec}})$ is different from that of the receiver $P_{rx_{elec}}$.

The Fair Propagation Problem: The propagation problem Π can be formulated as following: "Which is the propagation strategy the particles should follow, in order to distribute the work load in "fair" fashion and report as many messages as possible to the sink, during the K time-steps of network's operation?". Π implies the need of cooperation between the particles in order to share the data dissemination burden and propagate efficiently most of the messages without spending much energy. The issue of *balanced energy dissipation* must be taken into account as well. A protocol's ability of delivering most of the reporting messages is quantified by the *Success Rate* P_s. P_s is the fraction of the number r of the events successfully reported to the sink, over the total number of events generated g, i.e. $P_s = \frac{r}{g}$. Additionally we measure the *Final Total Energy* E_{fin}. Given a network of size N where each node i has E_i energy units left at the end of the simulation, $E_{fin} = \sum_{k=1}^{N} E_k$. To assess how energy balanced a protocol is, we calculate the *Standard Deviation of the Energy Left*, $\sigma_{E_{fin}}$. For every particle i having E_i units of energy, $\sigma_{E_{fin}} = \sqrt{E(E_i^2) - E(E_i)^2}$. Other approaches [12] characterize a protocol as energy balanced if no sensor dissipates more than $O(E(n)/n)$ energy, where n is the number of sensors and $E(n)$ the total amount of energy spent. We also measure the *Number of Alive Particles* H_a, i.e. the number of particles i that have, at the end of the simulation, energy levels above zero, i.e. $E_i > 0$. Finally we calculate how fast a protocol delivers its messages using the *Mean Latency* metric, $\overline{L}$. If an event ε belongs to the set of the successfully delivered events $\mathcal{E}_d$ and made it to the sink after l_ε rounds, then $\overline{L} = \frac{\sum_{\varepsilon \in \mathcal{E}_d} l_\varepsilon}{|\mathcal{E}_d|}$.

3   Implementation Issues

The implementation assumes that all the particles operate within discrete timesteps. This implies synchronization, although the latter is not necessary. The regular and the special "energy level update" messages can be delivered at any time and each particle is free to store and process them at the proper moment. We assume as well that the particles avoid message re-transmissions by keeping a record of the previously encountered ones. Also we make the assumption that an underlying MAC communication protocol exists dealing with the interference problem, as the *radiogram* protocol applied on the *Sun SPOT* motes [1]. Additionally we consider the *sleep/awake* scheme as an underlying subprotocol that can be applied on every classic routing protocol and thus we don't take its application into account in this study; all the particles are constantly awake. Each experiment is executed on the same hundred of similar *instances* and the results harvested are the average values from those taken on every single instance. The specifications utilized concern: the *network size* which is 10000 particles. The plane which is shaped as a unit diameter disk. The sink's position in the disk's center. Additionally the events are getting detected close to an arc which is a part of disk's circumference; the "event arc". The size of the "event arc's" subtending angle is 30°. The injection rate is 0.4 and the β parameter (express the length of PFR's "Front Creation" phase) is 2. The particles' communication radii and initial energy reserves are 0.025 length units and 5000 energy units respectively. Each experiment lasts 5000 rounds. The energy specifications are based on real data drawn from data-sheets of six well known micro-sensor transceiver architectures (**μAMPS-1**, **WINS**, **nRF905**, **MEDUSA II**, **TR1001** and **CC1100**). Here we report only the experiments regarding the μAMPS-1 architecture. Similar results hold for the other architectures as well. All these characteristics describe an average case scenario where the network works sufficiently for most of the time, but a significant fracture of the network gets depleted in order to evaluate protocols' robustness. More details regarding the experimental conditions can be found at the full version of this paper: http://ru1.cti.gr/aigaion/?TR=446.

4   Probabilistic Forwarding Protocol - PFR

The *Probabilistic Forwarding Protocol* (PFR) is a flat, distributed and multipath protocol and it favors transmissions only from those sensor particles that lie close to the *optimal line*; i.e. the line connecting the source of a message and its destination, the sink. It evolves in two phases:

The Front Creation Phase. This effect of this phase is to ensure that the data propagation process survives for an appropriately large period of time. It initiates when a new event $\mathcal{E}$ is detected by a particle p and $\mathcal{E}$'s message $m_{\mathcal{E}}$ is sent to p's neighbors. It lasts for β rounds and during them each particle receiving a message w it deterministically forwards it.

The Probabilistic Forwarding Phase. After the Front Creation phase data propagation is done in a probabilistic manner. When a particle v receives a new message m, it decides to propagate it with probability $p_{fwd} = \frac{\phi}{\pi}$. ϕ is the angle formed between the particle-source A of the message m, particle v and the sink S ($\phi = \widehat{AvS}$).

PFR's main drawback is its flooding behavior. This behavior is illustrated in Fig. 1(a). Flooding, favors the creation of multiple paths and thus the delivery probability is increased while the energy dissipation is increased as well.

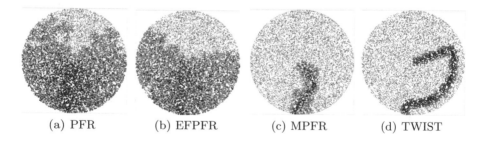

(a) PFR   (b) EFPFR   (c) MPFR   (d) TWIST

Fig. 1. Single message's propagation. The light grey spots stand for particles having received the message and the dark ones for nodes having received and transmitted it.

5   Designing Protocols

5.1   Energy-Fair Probabilistic Forwarding Protocol - EFPFR

EFPFR's goal is to limit message broadcasts and thus energy consumption, without blocking events from being reported to the sink. The idea of limiting transmissions was also utilized from different perspectives in [9]. EFPFR inherits both the Front Creation and Probabilistic Forwarding Phase from PFR, but it slightly modifies the latter one. It introduces the "local energy ranking" criterion which in combination with the original PFR's "angle" criterion, regulates the transmission decision process. A similar criterion was used in [7]. The *local energy ranking* criterion requires that each particle should have some knowledge about the energy reserve levels of its neighbors. Thus periodically all the particles are broadcasting information about their own energy level. Those special messages are the *energy update messages*, the special rounds are the *energy update rounds* and the number of rounds intervene between two consecutive energy update ones is called the *energy update period length*. During the energy update rounds, each particle receives its neighbors' energy information. Thus it can compare its own energy reserve with those of its neighbors, and assess whether it is locally "rich" or "poor", in terms of energy. If it is "rich", it concludes that it should take the initiative to forward its messages and relieve its "poor" neighbors. If it is "poor" it reduces the forwarding probability of its message since it "assumes" that there is another, "richer" neighbor to do so. Note that with high probability in a neighborhood all the particles possess more or less the same set of messages, as long as they have common neighbors to pass them the same messages. While in an energy update round, a sensor particle can still sense events, which will be processed during the next round of operation.

Fairness in EFPFR is expressed by the fact that in each neighborhood the burden of forwarding messages is undertaken by the "richest" particles relieving the "poor" ones. The energy distribution throughout the network may change in time. Since the energy update rounds occur periodically, the protocol adapts to the new environment. Summing up, during the EFPFR's Probabilistic Forwarding phase, a particle v forwards a message originating from particle A with probability:

$$p_{fwd} = \begin{cases} \frac{\phi}{\pi} \times \frac{E_v - E_{min} + 1}{E_{max} - E_{min} + 1} & , \text{ if } v \not\equiv A, \\ 1 & , \text{ if } v \equiv A. \end{cases}$$

E_v, E_{min} and E_{max} are respectively the energy level of particle v and the minimum and maximum energy levels in v's neighborhood including its own energy level, as they were reported during the last energy update round. The "+1" factor added at the numerator and the denominator, is normalizing the fraction in the case $E_{min} = E_{max}$. Note that the probability of broadcasting a message by the message's source particle is always 1. The only parameter left open is the energy update period length. A small period is keeping the energy information up to date but it causes frequent energy update rounds costing more energy. Our simulations showed that an energy update period of 80 rounds is optimal (details can be found in the full version of this paper). As it was expected and can be showed experimentally (see section 6), EFPFR improves PFR's performance because it saves energy truncating unnecessary message transmissions and some receptions. The message flooding is still present since only few transmissions in a dense network are enough to propagate data everywhere. This is depicted at Fig. 1(b).

5.2   Multi-phase Probabilistic Forwarding Protocol - MPFR

MPFR follows a different approach from EFPFR's one to achieve *fairness*. MPFR's concept is to forward messages not through the "optimal line", but rather through longer, crooked routes. This design choice works well on instances where all, or most of the events occur and get detected in a specified area. This assumption is reasonable, since in real applications a critical phenomenon is usually observed in specific regions and not throughout all the monitored space. MPFR's benefit is that energy dissipation per message is more balanced as the particles on the "optimal line" get some relief when the traffic is diverted away. In addition more potential propagations paths are created, increasing message's probability of delivery. On the other hand this technique affects the mean message delivery delay and increases the overall energy consumption.

MPFR calls PFR multiple times in order to form the indirect, crooked transmission path. In more detail, MPFR divides the network area using concentric rings, whose radii are integer multiplicates of the radius of the smaller one, also called *redirection radius* R_{rd}. The center of all these rings is the sink. Consequently the number of the rings is set by R_{rd}. Thereupon when a particle p detects an event $\mathcal{E}$, it creates the reporting message $m_{\mathcal{E}}$ and calculates its own distance d_p, from the sink. If $d_p \leq R_{rd}$ PFR is used to carry $m_{\mathcal{E}}$ directly to the sink. Else p picks the next *intermediate destination* of $m_{\mathcal{E}}$. Apparently R_{rd} and d_p set the number of PFR calls. Thus if $kR_{rd} < d_p \leq (k+1)R_{rd}$ and $k \geq 1$, PFR will be called $k+1$ times and $m_{\mathcal{E}}$ will be redirected k times. Picking the next destination particle p' is not trivial. p' is picked by p, uniformly at random from those particles lying close to the so called "choice arc". The original design choice suggests that the "choice arc" should be a hemicycle having as center p and whose radius would be equal to $d_p - (k-m) \cdot R_{rd}$. Additionally the bisector of this arc should coincide with the line connecting p and the sink. An example is illustrated on Fig. 2(a), where an event originally detected by particle A is reaching the sink, passing through B and C. The light grey line represents the message's routing, while the two dark arcs are the "choice arcs" of particles A and B. The problem is that sometimes a message is forced to backtrack and it gets blocked as the particles which had encountered it in the past ignore it. In the case showed on Fig. 2(a) this could happen if B's choice would be D instead of C. This "message blocking" phenomenon has a strong negative effect on protocol's performance. To cope with it we proposed two measures. The first is the *non-blocking* feature which declares that the particles should consider a message, when it is redirected, as different from the original one and not block it. The other one is the "periphery-targeting" mechanism. This concept changes the definition of the "choice arc".

It defines it to be part of the next outer concentric ring towards the sink and whose ends are the intersection points between the ring itself and the two tangent lines from particle p to the ring. Fig. 2(b) visualizes it. The goal of this strategy is to minimize the probability of message blocking because of backtracking, since the protocol always targets forwards.

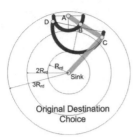 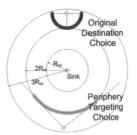

(a) A message from A propagates through B and C to the sink using the original destination choice method

(b) The original destination method's "choice arc" is dark, while the periphery targeting concept's one is light

Fig. 2. Design elements of the MPFR protocol

Our simulations indicate that despite of the correcting measures we introduced, MPFR's performance was still poor and specifically slightly worse than that of PFR. The main explanation is that MPFR is relying its success on the ability to bypass the direct route, from a message's source to the sink which is not possible because of the still present flooding behavior. The solution we turned to, was the to make the PFR's angle criterion stricter. To do so instead of forwarding a message with probability $\frac{\phi}{\pi}$ we forward it with probability $\left(\frac{\phi}{\pi}\right)^n$ where n is an integer number greater than 1. Fig. 1(c) depicts how MPFR propagates a message to the sink when it follows both the "non-blocking" and the "periphery targeting" concepts and when the propagation criterion is raised to the power of 5. The extensive experiments we conducted showed that the MPFR performs best when the criterion's power is 5, the redirection radius is large and when both the "non-blocking" and the "periphery targeting" features are applied. Details regarding these simulations and the overall design process of MPFR can be found at the full version of this paper.

5.3   Twisting Forwarding Protocol - TWIST

MPFR was engineered to enforce *fairness* by probabilistically deflecting data propagation away from the area close to the "optimal line". Its design aimed at indirectly achieve this goal and so its success was only partial. Using our experience on MPFR we designed protocol TWIST which aims exactly at directing data propagation away from greedy choices; the straight routes. TWIST considers that the network is deployed in a disc, at the center of which lies the sink.

TWIST runs in three phases. During the first one, the message is forwarded from the source particle E along the "optimal line" but only until a randomly selected point; the *twisting point* T. Then the second phase, called "twisting phase", is initiated and the message is forwarded along the perimeter of the circle having as center the sink and whose radius is equal to the distance between the sink and the twisting point. This twisting propagation is done clockwise or anticlockwise, and the covered arc's subtending angle is random (between 0 and π). The other end of the arc is called *direct forwarding point* D. Reaching this point the third phase is initiated and the message follows once again a straight route towards the sink S. The line, formed by the two straight lines of the first and the third phase and by the arc of the second phase, is called *propagation line*. An example of TWIST propagation is illustrated on Fig. 1(d).

This protocol is probabilistic since, for each message the positions of the twisting and direct forwarding points are random. Those positions are picked from the source-particle and they are recorded in the message. All the other decisions are deterministic. The reason is that we want to keep the three individual propagation routes "thin" enough to avoid flooding and spoil protocol's character. So a particle transmits a message, if and only if its distance from the propagation line is equal or shorter to its communication radius. The area around the propagation line where transmitting is permitted is called *propagation zone*. Details regarding TWIST's implementation are provided in the full version of this paper.

6   Comparative Study

Table 1 contains the performance measurements of the four protocols. As we see, EFPFR ameliorates PFR's success rate and this improvement is obvious to every single metric. Even the standard deviation of energy is smaller than that of PFR, implying a fairer distribution of the workload. The limitation of PFR's flooding behavior is obvious comparing Figures 3(a) and 3(b). Thus EFPFR indeed is improving PFR's performance, without compromises.

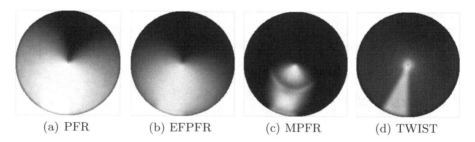

(a) PFR   (b) EFPFR   (c) MPFR   (d) TWIST

Fig. 3. The distribution of energy dissipation. Light painted areas host the most depleted particles, while the less depleted ones lie on the dark ones.

MPFR as shown in Table 1 succeeded in its mission. It achieved a success rate higher than that of EFPFR, it saved twice more energy than PFR and preserved alive much more particles than EFPFR did. Its standard deviation of energy was also much better than EFPFR's. From the other hand Fig. 3(c) shows that the key to MPFR's success is not that it enjoins *fairness* throughout the disk, but rather that it restrains flooding, and provides multiple delivery paths. TWIST finally seems achieving the best performance (97.5% in terms of success rate), keeps alive more particles, and the total amount of energy spared is comparable to that of MPFR. TWIST's success is the reduction of the energy's standard

Table 1. Comparative study of protocols PFR, EFPFR, MPFR and TWIST

Protocols	Parameters							
	Success Rate	Final Tot. Enrg.	Alive Particl.	Std. Dev. Energy	Mean Enrg.	Min Enrg.	Max Enrg.	Mean Latency
PFR	69.5%	20.227.374	7470	1845.47	2023	0	4862	29.03
EFPFR	81.1%	29.652.830	9199	1768.33	2965	0	4847	28.47
MPFR	88.9%	42.289.624	9824	1275.97	4229	0	4862	26.22
TWIST	97.5%	39.323.864	9895	1006.78	3932	0	4828	38.38

deviation (see Fig. 3(d)). As expected, the only drawback TWIST faces is that its message's delivery latency is significantly elongated.

7   Closing Remarks

Our future plan is to carry on the work on discovering the principles governing a good communication protocol for WSN's. We aim at extending our investigations on instances where obstacles exist, and in the same time enhance our experimental findings with analytical proofs. Finally we plan to take further our research on the mechanisms that ensuring *fairness* and we will work with another important virtue: *adaptability*.

References

1. Sun Small Programmable Object Technology (Sun SPOT) Developer's Guide, V3 "Purple" release, http://www.sunspotworld.com/docs/
2. Nikoletseas, S.E., Rolim, J.D.P. (eds.): ALGOSENSORS 2004. LNCS, vol. 3121. Springer, Heidelberg (2004)
3. Chatzigiannakis, I., Dimitriou, T., Nikoletseas, S.E., Spirakis, P.G.: A probabilistic algorithm for efficient and robust data propagation in wireless sensor networks. Ad Hoc Networks 4(5), 621–635 (2006)
4. Dimitriou, T., Kalis, A.. Efficient delivery of information in sensor networks using smart antennas. In: ALGOSENSORS [2], pp. 109–122
5. Falck, E., Floréen, P., Kaski, P., Kohonen, J., Orponen, P.: Balanced data gathering in energy-constrained sensor networks. In: ALGOSENSORS [2], pp. 59–70
6. Fekete, S.P., Kröller, A.: Topology and routing in sensor networks. In: Kutyłowski, M., Cichoń, J., Kubiak, P. (eds.) ALGOSENSORS 2007. LNCS, vol. 4837, pp. 6–15. Springer, Heidelberg (2008)
7. Hong, X., Gerla, M., Wang, H., Clare, L.: Load balanced, energy-aware communications for mars sensor networks. In: IEEE Aerospace, Bigsky (March 2002)
8. Karl, H., Willig, A.: Protocols and Architectures for Wireless Sensor Networks. John Wiley & Sons, Chichester (2005)
9. Leone, P., Moraru, L., Powell, O., Rolim, J.D.P.: Localization algorithm for wireless ad-hoc sensor networks with traffic overhead minimization by emission inhibition. In: Nikoletseas, S.E., Rolim, J.D.P. (eds.) ALGOSENSORS 2006. LNCS, vol. 4240, pp. 119–129. Springer, Heidelberg (2006)

10. Rangwala, S., Gummadi, R., Govindan, R., Psounis, K.: Interference-aware fair rate control in wireless sensor networks. SIGCOMM Comput. Commun. Rev. 36(4), 63–74 (2006)
11. Rao, A., Ratnasamy, S., Papadimitriou, C., Shenker, S., Stoica, I.: Geographic routing without location information. In: 9th ACM/IEEE Annual International Conference on Mobile Computing (MOBICOM 2003), pp. 96–108 (2003)
12. Singh, M., Prasanna, V.K.: Energy-optimal and energy-balanced sorting in a single-hop wireless sensor network. In: PerCom, pp. 50–59 (2003)

Simple Robots in Polygonal Environments: A Hierarchy

Jan Brunner[1], Matúš Mihalák[1], Subhash Suri[2,*],
Elias Vicari[1,**], and Peter Widmayer[1,**]

[1] Department of Computer Science, ETH Zurich, Zurich, Switzerland
jabrunne@student.ethz.ch, {mmihalak,vicariel,widmayer}@inf.ethz.ch
[2] Department of Computer Science, University of California, Santa Barbara, USA
suri@cs.ucsb.edu

Abstract. With the current progress in robot technology and related areas, sophisticated moving and sensing capabilities are at hand to design robots capable of solving seemingly complex tasks. With the aim of understanding the limitations of such capabilities, swarms of simple and cheap robots play an increasingly important role. Their advantages are, among others, the cost, reusability, and fault-tolerance. While it can be expected that for a variety of problems a wealth of robot models are proposed, it is rather unfortunate that almost all proposals fail to point out their assumptions explicitly and clearly. This is problematic because seemingly small changes in the models can lead to significant differences in the capabilities of the robots. Hence, a clean assessment of the "power of robot models" is dearly needed, not only in absolute terms, but also relative to each other. We make a step in this direction by explaining for a set of elementary sensing devices which of these devices (alone and in combination) enable a robot to solve which problems. This not only leads to a natural relation (and hierarchy) of power between robot models that supports a more systematic design, but also exhibits surprising connections and equivalences. For example, one of the derived relations between the robot models implies that a very simple robot (that cannot measure distances) moving inside a simple polygon can find a shortest path between two vertices by means of a sensor that detects for an angle at a vertex of the polygon whether it is convex. We give an explicit algorithm which allows the robot to find a shortest path.

1 Introduction

Nowadays, rapid technological innovation gives rise to new hopes and exciting possibilities for microrobots. For instance, camera sensors continuously become

* The author gratefully acknowledges the support of the National Science Foundation under research grants CNS-0626954 and CCF-0514738.
** Work partially supported by the National Competence Center in Research on Mobile Information and Communication Systems NCCR-MICS, a center supported by the Swiss National Science Foundation under grant number 5005 − 67322.

S. Fekete (Ed.): ALGOSENSORS 2008, LNCS 5389, pp. 111–124, 2008.

smaller, cheaper and provide images of higher quality. It is natural for robotics engineers to build sophisticated robots that use many of the available features.

It is, however, not always clear that the more complex information gained from more sophisticated sensors is what a robot needs in order to solve a task at hand. For instance, the project SToMP (Sensors, Topology, and Minimalist Planning) of the University of Illinois investigates questions of this nature. The current understanding of robotics is not united in answering the question of which sensor is better for certain tasks. Giving all possible sensors/features to the robot may prevent scientists from asking such questions, but it does not come for free, and one has to justify the increased cost of building, memory usage, energy consumption, etc. Recently, Seth Teller gave a Plenary talk at IPSN on Darpa's Self-Driving car Grand Challenge [1]. The car they built for the challenge was completely self-driving in a real-world traffic, with other robot cars, and several stunt drivers. The interesting aspect that Seth Teller pointed out was the "sensory overload". Even with 40 CPUs packed in the trunk of the car, they could not keep up with all the information that the sensors were feeding, and it was also clear that most of the sensing information was not being used. While in this paper we stick with a purely theoretical point of view, this example shows that a fundamental research that identifies the "necessary and sufficient" sensing information for a task (such as driving) would be very valuable. It is therefore an important research question to understand the limitations and strengths of available features/sensors.

We take a step in this direction and try to understand what is the information (gained via sensors) that a robot *necessarily* needs to solve a certain task. We compare several sensors within the framework of simple robots in polygonal environments. Such sensors can be e.g. the capability to measure distances between visible vertices, or the capability to measure angles between visible vertices. We compare the resulting robot models with the aim to determine their computational strength. At this stage of our investigation we do not want to compare how effective the robots may be in dealing with a massive amount of data. Instead we base our comparison on the notions of "possibility" – we are only interested whether the considered robots can solve the same task, and we do not compare how much time, memory, energy, etc. the robots need to complete the task.

To give a concrete example, consider a polygonal maze with one exit and no holes, where every wall has a different color. There are three different simple robots placed in the maze. All robots can see the vertices and the walls of the polygonal maze. Robot A can additionally measure the distance between its position and a visible vertex, robot B can see the color of the wall, and knows the order of the colors as they appear on the boundary, and robot C can decide whether an angle between two visible vertices and its position is convex. These robots have seemingly very different level of sophistication. The task for the robots is to find the exit. This is of course an easy task as a robot can walk along the boundary and it eventually finds the exit. In this paper we prove that the three robots can do more: they can determine a Euclidean shortest path between their position and the exit of the maze. Even though this sounds plausible for robot A, we think it is very surprising for the other robots that have no distance measuring capability. This motivates the need to understand the true strengths (and weaknesses) of idealized robots.

On the practical side and with the important growth of *wireless sensor networks* [2], a group of simple and small robots also offers an attractive and scalable architecture for large-scale collaborative deployment and exploration of unknown environments.

The remainder of the paper is organized as follows. We first relate our work and models to the existing literature. We then define the simple robot model we are dealing with, together with the class of sensors that we are going to consider. We define the relation *stronger* which is used in the subsequent sections to compare the resulting robot models. At the end we resolve the problem from our motivating example, namely, we show how a robot C, which can recognize whether an angle between two given visible vertices is convex, can find a shortest path in a polygon. Furthermore we show that robot B (which can "see the colors") is "equivalent" to robot C and thus it can solve the maze problem, too.

Related Work. In this paper we study idealized robot models which are similar to the ones described in [3,4,5,6]. This class of robots has proved to be useful as the robots are able to solve involved geometrical problems like the Art Gallery Problem [3] or Pursuit-Evasion [5] despite their severe sensing limitations.

In the literature there have been other efforts to introduce a hierarchical description of robot models – see for example [7], which summarizes some of the recent work. The idea of understanding and comparing robots' sensor-equipment and the underlying strength of resulting robots is not new. Donald [8] defines and studies among the first general concepts of hierarchies among robots. His work deals, analogously to our, with the problem of determining the information requirement of a robot to solve a task. He also introduces a general concept for comparing different robot models. The present analysis differentiates from the previous work in two main aspects. Firstly, instead of considering the universe of robots we focus on a specific class of robots. Each robot that we consider is derived from a single basic robot by equipping it with new devices of different sophistication. Such a restriction allows more precise and formal reasoning. Further, along with a hierarchy which represents the *relative* strengths of the robots, we study the *absolute* strengths of the robots. To do so, we try to understand which information about the environment can be extracted by each robot. In the considered model we also apply results of computational geometry and get to the information-theoretical core of sensors.

2   Definitions and Models

Polygons and Visibility. In this paper we deal exclusively with *simply-connected polygons*. A simply-connected polygon P is the compact region of the plane bounded by a closed, non-selfintersecting polygonal curve, called the *boundary* (or *boundary curve*) of P. We denote by $V(P) = \{v_0, \ldots, v_{n-1}\}$ the set of vertices of the boundary curve, which we assume are ordered counterclockwise (ccw). Every vertex v_i, $0 \leq i \leq n - 1$, is a point in the Euclidean plane. We assume, for simplicity of exposition only, that no three vertices lie on a line. Let $E(P) = \{e_0, \ldots, e_{n-1}\}$ be the edge-set of a polygon P, where an edge is the *open* segment $e_i = (v_i, v_{i+1})$, $i = 0, \ldots, n - 1$. All arithmetic operations on the indices of the vertices are to be understood modulo n. Let $\mathcal{P}_n$ be the set of all simply-connected polygons on n vertices in the Euclidean plane, and let $\mathcal{P}$ be the set of all simply-connected polygons, i.e., $\mathcal{P} = \bigcup_{n \geq 3} \mathcal{P}_n$. Given two points p_1 and p_2 in the polygon, we denote by $p_1 p_2$ the line segment between the two points. We say that p_2 is *visible* to p_1 in P if $p_1 p_2 \cap P = p_1 p_2$ (note that the relation *visible* is symmetric). Two (mutually) visible vertices v_i and v_j, $j \notin \{i - 1, i + 1\}$ form a *diagonal* of P.

The *visibility graph* of a polygon P is the (labelled) graph $G = (V, E)$ with $V = V(P)$ where two vertices are adjacent iff they are (mutually) visible in P. The *vertex-edge visibility graph* is a bipartite graph $G' = (V', E')$ with $V' = V(P) \sqcup E(P)$ where two vertices $v \in V(P)$ and $e \in E(P)$ are adjacent in G iff v is visible to at least a point of the edge e in P.

A Robot Hierarchy. The set of problems that we are interested in is defined by the universe of mappings $f : \mathcal{P} \times \mathcal{I} \to \mathcal{O}$, where $\mathcal{I}$ denotes the set of inputs (additional to the polygon) and $\mathcal{O}$ the set of outputs, respectively. Instead of using a more classical Turing machine as a computational model to solve the aforementioned problems, we use robots. An *instance* of a problem f is defined by a pair (P, I), where $P \in \mathcal{P}$ and $I \in \mathcal{I}$. Given a problem f, a robot is said to *be able to solve* f if, given any instance (P, I) of the problem f, a robot R that moves according to its deterministic specifications in the polygon P returns a solution $O \in \mathcal{O}$ which (1) is independent of the vertex where the robot is initially located, (2) fulfills $O = f(P, I)$, and (3) needs a finite number of movements of the robot. For example, consider the problem of determining the euclidean shortest path between two vertices s and t of a polygon P (on n vertices), which lies entirely inside P. We set $\mathcal{I} = \mathbb{N}^2$, which represents the choice of s and t, $\mathcal{O} = \bigcup_{P \in \mathcal{P}}\{\text{set of all polygonal paths in } P\}$ and f is the map that given a polygon P and two vertices of P returns a shortest path inside P between those two vertices.

In this paper we consider a robot as a computational model for problems in polygons. Robots are characterized by given features and limitations, which define a *robot model*[1] – we denote by $\mathcal{R}$ the set of all robot models that operate in polygons. For a robot model $R \in \mathcal{R}$ we want to identify the set of problems that a robot can solve. Accordingly we define a partial order on the set of robot models. We say that a robot R_1 is *at least as strong* as the robot R_2 – denoted $R_2 \trianglelefteq R_1$ – if R_1 can solve all the problems that R_2 can solve. We say that R_1 is

---

[1] When this is not misleading, we often do not distinguish the concepts of a robot and a robot model.

strictly stronger $(R_2 \lhd R_1)$ if additionally there is a problem that R_1 can solve but R_2 cannot. R_1 and R_2 are *equivalent,* if $R_1 \unlhd R_2$ and $R_2 \unlhd R_1$.

Robot Models. All considered robot models of this paper are extensions of a basic robot model – the *simple combinatorial robot* (the definition follows), which provides elementary motion and vision capabilities. We model additional sensing capabilities by *devices* which can be "mounted" onto the basic robot. Any set of devices derives a new robot model from the basic robot. We define the considered devices after the introduction of the basic robot.

The *simple combinatorial robot* has very basic sensing and motion capabilities. We assume that the robot is placed on a vertex of a polygon P. The robot possesses a sense of *handedness*. It can locally distinguish its immediate "left" and "right": a robot at vertex v_i distinguishes v_{i-1} as the "left" neighbor and v_{i+1} as the "right" neighbor. In general, a robot at vertex v gets the following information about the environment. The sensory system scans the surroundings counterclockwise from the right neighbor of v. This produces an ordered list of visible vertices (including v itself), starting with the right neighbor of v and ending with the left neighbor and the vertex v. The vertices are visually indistinguishable to the robot (they are unlabelled). Hence, the robot cannot in general recognize a vertex, that is, cannot detect whether a vertex seen from one location is the same as another vertex seen from another location. The sensing process produces the so called *combinatorial visibility vector* (cvv for short) of the vertex v. The cvv is a binary vector encoding the presence of edges between the visible vertices. The ith component of cvv is 1 if between the $(i-1)$st and ith visible vertex there is a polygonal edge, and it is 0 if there is a diagonal (see Fig. 2 for an example).

The motion ability is likewise very limited. The robot picks a destination among the visible vertices and moves on a straight line – along a polygonal edge or diagonal – until the destination is reached. Thus, if the robot does not move, it stands on a vertex of P. During the motion no sensing is possible.

Our focus is to study the possibility issues of the simple robots equipped with various devices. Therefore we assume that the robot has an unbounded computational power and an infinitely large memory with arbitrary precision. We note that the simple combinatorial robot has the sensory and motion systems similar to the robots studied in [3].

We consider the following, some of the most common devices in robotics.

Pebble. A *pebble* is a device that is used for marking the vertices. The robot can drop the pebble at the vertex of the robot's position, and can recollect it again for further use. If the vertex with the pebble is visible to the robot's position, the robot sees the vertex as marked, and distinguishes this from the other vertices. The most important implication of this is the following. Suppose that the robot is on the vertex v_i, leaves the pebble at this vertex and moves to the vertex v_j. There, at vertex v_j, the robot can determine the relative position of v_i in the robot's cvv.

Angle-Measuring Device, Length-Measuring Device. Suppose that a robot is on vertex v_i and that two vertices v_j and v_k are visible from v_i.

Let us assume, w.l.o.g., the positions of v_j and v_k in the cvv of v_i are j' and k', respectively, $j' < k'$. Upon a request for the angle between the visible vertices j' and k' of the cvv, the *angle-measuring device* returns the exact angle $\angle v_j v_i v_k$, i.e. the angle that "lies" entirely inside the polygon. Analogously, the *length-measuring device* measures the distance between the robot's current location and a chosen visible vertex.

Reflexity Detector. The *reflexity detector* is similar to the angle-measuring device but instead of providing the size of the angle it merely decides whether the angle is *convex* or *reflex*, i.e., whether the angle is smaller, respectively bigger, than 180 degrees.

Compass. The *north-direction* is a consistent reference direction in the polygon and is parallel to the y-axis in the coordinate system of the polygon. The *compass* enables the robot to measure the angle formed by the north-direction, the robot's current location and a chosen visible vertex.

Oracle. An *oracle* Ω is a (possibly non-deterministic) "black-box" device that can answer pre-specified *questions* posed by the robot. An oracle has perfect knowledge of the universe (the polygon, the robot's position, history of movements, etc.). Note that the other devices can all be seen as oracles.

The *labeller oracle* Ω_l answers the following type of questions. The query to the oracle is a component i of the cvv of v. If the component is 1 then the oracle reports the (global) label k of the corresponding edge e_k, i.e., the edge between the $(i-1)$st and ith visible vertex from v. If the component is 0, then the oracle reports the label of the edge that is *partially visible* behind the corresponding diagonal. This is the first edge that is intersected by the line that emanates from the vertex v and passes through the $(i-1)$st visible vertex of v (see Fig. 1). Note that the line that emanates from v and passes through the ith visible vertex intersects the same edge.

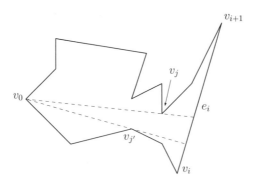

Fig. 1. Partially visible edge $e_i = (v_i, v_{i+1})$ from vertex v_0

We investigate some interesting combination of devices. In the following table we name the considered robot models and the devices that specify them.

Explorer	Pebble
Reflexer	Pebble, reflexity detector[2]
Labeller	Pebble, labeller oracle Ω_l [3]
Surveyor	Pebble, angle-measuring device

3   Combinatorial Structures of a Polygon

All robots of our interest have limitations in the motion and sensing capabilities, but no limits on their computational power or memory size. In that respect an ability of a robot to solve a problem is closely related to the ability of building a good representation of the environment. The goal of this section is to characterize, where possible, the information that a robot can extract from a polygon – its *map*. To do so, we compare according to the aforementioned complexity hierarchy the robot models with the so called *delphic robot*[4]. The delphic robot has the conceptually unbounded computational unit and the memory of the simple combinatorial robot but it can neither sense the environment nor move. For the comparison we characterize the delphic robot by a specific oracle Ω. The nature of the oracle will help us understand the type of information that the compared robot can extract from the polygon.

Theorem 1. *The explorer is equivalent to the delphic robot with the oracle that returns the visibility graph of the polygon.*

Proof. [3] shows that the explorer can build the visibility graph of a polygon. For the other direction, note that the cvv of every vertex can be read from the visibility graph, and also the information about a placed pebble (since the visibility graph is labelled). ☐

Using the same arguments, one can show that a simple robot with more pebbles is equivalent to the explorer. In a similar way one can also prove the following theorem.

Theorem 2. *The labeller is equivalent to the delphic robot with the oracle that returns the visibility graph and the vertex-edge visibility graph of the polygon.*

[9] shows that if no three vertices of the polygon are on a line, the visibility graph of a polygon can be computed from the vertex-edge visibility graph. This is not true in general otherwise.

To study the strengths of the surveyor, which seems to be a very strong robot, we introduce an oracle Ω_s. Let $\mathcal{A}$ be the set of *orientation-preserving similarity transformations*[5] of the plane. When queried, Ω_s returns the sequence

---

[2] Reflexer is the robot C of the introduction.

[3] Using colors to label edges, we see that labeller is the robot B of the introduction.

[4] Delphi is known for the oracle at the sanctuary that became dedicated to Apollo during the classical period.

[5] An orientation-preserving similarity transformation is a point-to-point mapping of the plane which is a composition of scalings and rotations followed by a translation.

$(v'_0, \ldots, v'_{n-1})$, where v'_i is the image (i.e. the coordinates) of the vertex v_i according to a mapping $m \in \mathcal{A}$, where m is fixed (independent of the query) and unknown to the robot.

Theorem 3. *The surveyor is equivalent to the delphic robot that is equipped with the oracle Ω_s.*

Proof. First observe that the surveyor can build the visibility graph of the polygon because it is an extension of the explorer and hence can easily navigate in the polygon P (move from a desired vertex to another).

Let the delphic robot have an instance of the oracle Ω_s that uses a map $m \in \mathcal{A}$. Consider now the surveyor. Pick a triangulation T of P and take an edge e of T which corresponds to an edge or a diagonal of P with endpoints u and v. The robot assigns arbitrary (distinct) coordinates to u and v to start building a representation of P. Take a vertex w that induces a triangle with e in T. The robot can measure perfectly the angles $\angle(u, v, w), \angle(v, w, u), \angle(w, u, v)$. This leaves exactly two points of the plane where w can be placed to represents correctly the angles of P. The handedness of the robot cuts down the number of options to one. It is easy to see that by further following the triangulation T the remaining vertices of P are placed uniquely. Hence the coordinates of the representation of P of the robot are equal to the image of the true coordinates of P under a (unknown) map $m' \in \mathcal{A}$. Note that this map is fully specified by the coordinates of the vertices u, v. This shows that the surveyor can simulate the oracle Ω_s. For the other direction observe that the angles that the delphic robot measures in the representation generated by the oracle are equal to the respective angles measured by the surveyor. $\qquad \square$

Suppose that we enrich the capabilities of the surveyor. In addition to the angle-measuring device, we endow the surveyor with either a length-measuring device or a compass (or both). By literally repeating the previous proof, we can easily see that such a robot builds a more accurate representation of the polygon. In other words, the enriched surveyor is equivalent to the delphic robot endowed with the same oracle Ω_s, but the set of geometric transformations which may be used by the oracle is smaller – the length-measuring device prevents the oracle to pick a transformation that involves a scaling while the compass prevents rotations. By again arguing with a triangulation of the polygon, it becomes evident that the length-measuring device can simulate the angle-measuring device. Obviously, the opposite direction does not hold.

We note that the angle-measuring device can be simulated by the compass and that the pebble can be simulated by the combination of the compass and the length-measuring device. To see the latter, notice that a robot equipped with the compass and the length-measuring device can simulate a polar coordinate system and hence give coordinates to the vertices that coincide with the true coordinates up to a translation. If coordinates are given, the robot can operate with labelled features and hence a pebble can be easily simulated.

4   A Robot Hierarchy

In this section we want to precisely investigate the relationships between the different robots. Trivially, "simple combinatorial robot $\trianglelefteq$ explorer".

Theorem 4. *The explorer is strictly stronger than the simple combinatorial robot.*

Proof. We show that the simple combinatorial robot cannot solve a problem that is trivial for the explorer: count the vertices of a polygon. Consider the two polygons depicted in Fig. 2. The polygon P_1 is only partially drawn. Easy geometric arguments show that P_1 can be extended to an arbitrary size with the property that vertices with cvv $(1, 0, 0, 1)$ and $(1, 1, 0, 1, 1, 0, 1, 1)$ alternate along the boundary (on both sides). If P_1 is made large enough, then its middle part, i.e. the portion of the polygon specified by all the vertices except for those with maximal and minimal x-coordinate, "looks like" the polygon P_2. Let us make this precise. The middle part of P_1 and of P_2 are composed by vertices of two different cvv's: $c_1 = (1, 0, 0, 1)$ and $c_2 = (1, 1, 0, 1, 1, 0, 1, 1)$. Moreover, take two vertices $v \in V(P_1)$ and $w \in V(P_2)$ with the same cvv. If we move to the ith visible vertex in both cases, we get to two vertices v', w', respectively, such that again $\mathrm{cvv}(v') = \mathrm{cvv}(w')$ holds. Provided that P_1 is large enough and v is chosen "in the middle" of P_1, this observation can be repeated a finite number of times.

Suppose that there is an algorithm $\mathcal{A}$ for a simple combinatorial robot to count the vertices of a polygon. Let the robot sitting on a vertex of type c_i, $i = 1, 2$, of P_2 execute $\mathcal{A}$. The robot will report 10 after moving a finite number of times – say t times. By the above argument, if we place the robot on a middle vertex of

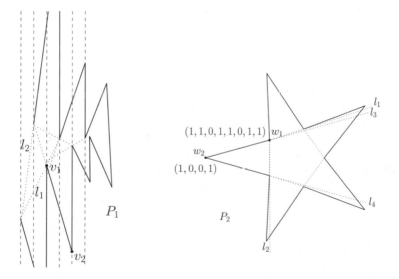

Fig. 2. P_1 and P_2 are the polygons used in the proof of Theorem 4

type c_i of P_1, it will visit the same sequence of vertices as in the execution in P_2, because it makes decisions only according to the sequence of cvv 's of the visited vertices. Thus it reports 10 after exactly t steps – this is a contradiction.   □

Theorem 5. *We have "explorer $\lhd$ labeller".*

Labeller is obviously at least as strong as the explorer. To see that it is strictly stronger take for instance the problem of deciding if a given vertex is convex or reflex. [3] shows that the explorer cannot solve this problem in general, whereas [9] shows that the vertex-edge visibility provides this information.

The next result is more surprising.

Theorem 6. *Labeller and reflexer are equivalent.*

Proof. As both robots are at least as strong as the explorer, both can build the visibility graph of the polygon.

We first prove that "labeller $\unlhd$ reflexer". To do so, we show that reflexer can use its capabilities to simulate labeller's oracle, i.e., we want to show that reflexer can identify the endpoints of every (partially) visible edge of the polygon. This is obvious if an edge e_i is *totally visible*, i.e., if both the vertices v_i, v_{i+1} are visible from the robot's current position – w.l.o.g. v_0. In this case the reflexer sees the endpoints, too, and thus can read the labels of the vertices, and hence the label of the edge, from the visibility graph.

Thus we concentrate on the case where the reflexer sees only a portion of the edge e_i. The edge e_i thus corresponds to a 0 (a diagonal) in the robot's cvv – let $v_j v_{j'}$ be this diagonal (the robot knows the labels j and j' from the visibility graph). The visible part of e_i is determined by the cone defined by the lines v_0, v_j and $v_0, v_{j'}$. See Fig. 1 for illustration. The robot first checks whether it sees one endpoint of e_i, i.e., whether $v_i = v_j$ or $v_{i+1} = v_{j'}$. The case $v_i = v_j$ happens iff the vertex v_{j+1} lies left of the line v_0, v_j, i.e., iff the angle $\angle v_0 v_j v_{j+1}$ is convex. Similarly, $v_{i+1} = v_{j'}$ happens iff the vertex $v_{j'-1}$ lies right of the line $v_0, v_{j'}$, i.e., iff the angle $\angle v_{j'-1} v_{j'} v_0$ is convex. In such a case reflexer can easily identify the labels of the endpoints of e_i.

It remains to consider the case when the robot does not see v_i and v_{i+1} from v_0. Thus, v_i lies to the right of the line v_0, v_j and v_{i+1} lies to the left of the line $v_0, v_{j'}$ (see Fig. 1). Observe that this is the only visible edge between v_j and $v_{j'}$ with this property. Thus, it follows trivially, e_i is the only visible edge which has one endpoint to the right of v_0, v_j and the second endpoint to the left of v_0, v_j. The robot then moves to v_j and checks whether it sees from v_j an edge where one endpoint v_k is to the right of v_0, v_j (again, this can be done by deciding whether the angle $\angle v_0 v_j v_k$ is reflex) and the second endpoint v_{k+1} is to the left of v_0, v_j. If it cannot find such an edge, the edge e_i (which is still partially visible from v_j; actually, a bigger portion is visible) forms the background of some diagonal in the robot's cvv. It is easy to identify which diagonal it is – the only diagonal formed by vertices v_k and $v_{k'}$ for which v_k lies to the right of the line v_0, v_j and $v_{k'}$ lies to the left of the line v_0, v_j. We have been in our analysis in this situation already – the robot now checks whether it sees one endpoint of

e_i, i.e., whether $v_i = v_k$ or $v_{i+1} = v_{k'}$. If yes, we are done, otherwise we proceed recursively: the edge e_i has its endpoints on the two sides of the cone formed by the lines v_j, v_k and $v_j, v_{k'}$, and it is the only edge between v_k and $v_{k'}$ with this property. Thus, the robot can move to v_k and perform the whole procedure again. This recursive approach has to stop, as the distance between k and k' is getting smaller in every step, and eventually $v_k = v_i$.

We now prove that "reflexer $\trianglelefteq$ labeller". Recall that by Theorem 2, labeller is already able to decide the type of an angle specified by three consecutive vertices. Consider the angle $\vartheta = \angle v_i v_j v_k$ and for simplicity assume that $1 \leq i < j < k \leq n$. ϑ is convex iff one of the edges $e_j, e_{j+1}, \ldots, e_{k-1}$ are partially visible to v_i. To see this, assume first that $k - j = 1$ and that ϑ is reflex. Then v_i cannot see e_j because it is hidden by v_j. Conversely suppose that ϑ is convex. Then by the assumption that the boundary curve is non-selfintersecting, v_i must be able to see a portion of e_j. The general claim follows by noting that v_i can see a portion of the diagonal $v_j v_k$ if and only if it can see a portion of one of the claimed edges. $\qquad\square$

5   Equivalent Robots: An Example

O'Rourke and Streinu [9] showed that the vertex-edge visibility graph provides enough information to determine the Euclidean shortest path between two points in a polygon (and also a shortest path tree from a given vertex). In the previous sections we have proved that reflexer is equivalent to labeller and that labeller is equivalent to the delphic robot that has access to the vertex-edge visibility graph. Hence we know that reflexer can compute a shortest path between two vertices of the polygon as well. A way for reflexer to do this is to follow the reduction to labeller and simulate the labeller's algorithm. This may not always lead to an intuitive algorithm for a problem. In this section we solve the shortest-path problem by reflexer directly, and present an algorithm that exploits naturally reflexer's features – deciding the convexity of any angle induced by two visible vertices and the robot's position.

To this end, we exploit the following structural theorem about paths in polygons. A *polygonal* path is a path induced by a sequence $l_1, l_2, \ldots, l_k$ of points in the plane – it starts at a vertex l_1 and always connects by a straight line to the next vertex of the sequence. We look only at polygonal paths that are entirely included in the polygon P and with $\{l_1, \ldots, l_k\} \subset V(P)$. An *internal angle* of a polygonal path at vertex l_i, $1 < i < k$, is the angle between the lines $l_i l_{i-1}$ and $l_i l_{i+1}$, that lies entirely inside P.

Theorem 7. *Let P be a simply-connected polygon and $s, t \in V(P)$. Then there is a unique polygonal $s - t$-path that turns at vertices of the polygon for which every internal angle is reflex. This path is the unique shortest $s - t$-path in P.*

Lee and Preparata [10] showed that the shortest $s - t$-path in a polygon is unique and has the claimed property. Nevertheless, we present a new proof because it is slightly simpler and additionally shows that no other $s - t$-path that turns at vertices of the polygon with reflex internal angles exists.

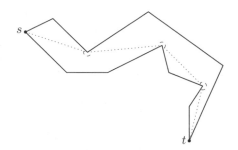

Fig. 3. The picture depicts a polygon with a shortest $s - t$-path. Note that all internal angles are reflex.

Proof. Figure 3 illustrates the situation of the theorem. It has already been proved in the literature that a shortest path between any two points in a polygon is a polygonal path that turns at vertices of the polygon [11]. Furthermore, it is not difficult to see that every shortest path has only reflex internal angles (i.e., angles bigger than 180 degrees): suppose that an internal angle $\angle l_{i-1} l_i l_{i+1}$ of a shortest (polygonal) path is convex. Then the general position assumption of the polygon implies that l_i can be moved slightly in the direction of the bisector of the angle $\angle l_{i-1} l_i l_{i+1}$ such that the new path does not cross the polygonal boundary. Obviously the newly created path is shorter (see Fig. 4), which contradicts our assumption. A more careful analysis shows that the general position assumption is not necessary (we omit the analysis here due to space limit).

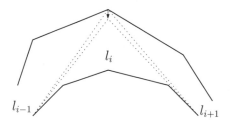

Fig. 4. The shortest path in a simple polygon has all its internal angles reflex

We now show that there is a unique polygonal $s - t$-path with all internal angles being reflex. This then shows that it has to be a shortest $s - t$-path.

Suppose for contradiction that we can find two distinct polygonal $s - t$-paths L_1, L_2 such that all their internal angles are reflex. Let p be the first vertex on L_1 from which the two paths differ, and let q be the first point on L_1 after p, where the two paths meet again (notice that q does not have to be a vertex of P). Let L_1' and L_2' be the induced sub-paths of L_1 and L_2, respectively, between p and q. Observe that L_1', L_2' induce a closed curve C. Observe also, that the region enclosed by C is completely inside the polygon.

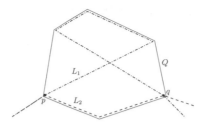

Fig. 5. The picture depicts the situation of the proof of Theorem 7. For simplicity the polygon P has been omitted.

Let Q be the polygon which is defined by the convex hull of the nodes of C. Note that Q has at least three vertices and that all vertices other than p and q are vertices of P. The situation is depicted in Fig. 5. As L_1' and L_2' lie in Q, the internal angle of any vertex of L_1' or L_2' is at most the respective angle of the polygon Q. However, the angles of Q are all convex, so every internal angle of a vertex of L_1' and L_2' must be convex. Thus there is a vertex w of a path L_1' or L_2' not equal to p and q such that the internal angle of w is convex. This vertex induces the same convex angle in the whole path (L_1 or L_2), which means there is a convex internal angle in one of those paths, a contradiction.   □

The new insight of Theorem 7 is that a polygonal $s - t$-path that turns only at polygonal vertices with reflex internal angles is unique. To find a shortest $s - t$-path in a polygon P, the reflexer can just find the unique polygonal $s - t$-path with only reflex internal angles. We describe an algorithm for reflexer that builds a shortest-path tree of P rooted at s. The robot starts at s and connects s with all visible vertices to begin the construction of the shortest-path tree T. The robot then proceeds iteratively: it goes to every leaf l of T and for every visible vertex v from the leaf the robot checks whether the angle between the predecessor of the leaf and v is reflex. If yes, it connects v to l in T. The following invariant follows easily from Theorem 7 and is maintained at any time of the algorithm: (i) The path in T from s to any vertex of T is the shortest path between the vertices in P, (ii) and T is a tree. It is not difficult to see that at the end T contains all vertices of P, and thus T is a shortest-path tree of polygon P.

The surprising fact is that to determine the Euclidean shortest path in a polygon we do not measure lengths. We remark that in multiply-connected polygons paths composed by reflex vertices are not unique in general (and have not the same length). Theorem 7 settles the intriguing example of the introduction.

As discussed above, the explorer is not able to recognize whether a vertex is convex or reflex. Subsequently the explorer is not capable in general of determining the Euclidean shortest path between two vertices in a polygon. If an algorithm for this problem would exist, then the explorer could easily exploit it to determine the type of the vertex v_i: v_i is reflex if and only if the shortest path between v_{i-1} and v_{i+1} goes through v_i.

Corollary 1. *The explorer cannot determine in general the Euclidean shortest path between two vertices of a polygon.*

References

1. Teller, S.: Development of a self-driving car as a mobile sensing platform. In: CPS Week Keynote Presentation at the International Conference on Information Processing in Sensor Networks (IPSN) (April 2008)
2. Pottie, G.J., Kaiser, W.J.: Wireless integrated network sensors. Commun. ACM 43(5), 51–58 (2000)
3. Suri, S., Vicari, E., Widmayer, P.: Simple robots with minimal sensing: From local visibility to global geometry. In: Proceedings of the Twenty-Second National Conference on Artificial Intelligence and the Nineteenth Innovative Applications of Artificial Intelligence Conference. AAAI Press, Menlo Park; Extended Version as ETH Technical Report 547 - Computer Science Department, pp. 1114–1120 (2007)
4. Gfeller, B., Mihalák, M., Suri, S., Vicari, E., Widmayer, P.: Counting targets with mobile sensors in an unknown environment. In: Kutyłowski, M., Cichoń, J., Kubiak, P. (eds.) ALGOSENSORS 2007. LNCS, vol. 4837, pp. 32–45. Springer, Heidelberg (2008)
5. Yershova, A., Tovar, B., Ghrist, R., LaValle, S.M.: Bitbots: Simple robots solving complex tasks. In: Proceedings of the Twentieth National Conference on Artificial Intelligence and the Seventeenth Innovative Applications of Artificial Intelligence Conference, pp. 1336–1341 (2005)
6. Ganguli, A., Cortes, J., Bullo, F.: Distributed deployment of asynchronous guards in art galleries. In: Proceedings of the American Control Conference, pp. 1416–1421 (June 2006)
7. O'Kane, J.M., LaValle, S.M.: Dominance and equivalence for sensor-based agents. In: Proceedings of the Twenty-Second National Conference on Artificial Intelligence and the Nineteenth Innovative Applications of Artificial Intelligence Conference, pp. 1655–1658. AAAI Press, Menlo Park (2007)
8. Donald, B.R.: On information invariants in robotics. Artificial Intelligence 72(1-2), 217–304 (1995)
9. O'Rourke, J., Streinu, I.: The vertex-edge visibility graph of a polygon. Computational Geometry: Theory and Applications 10(2), 105–120 (1998)
10. Lee, D.T., Preparata, F.P.: Euclidean shortest paths in the presence of rectilinear barriers. Networks 11, 285–304 (1984)
11. Mitchell, S.B.: Geometric shortest paths and network optimization. In: Handbook of Computational Geometry, pp. 633–701 (2000)

Deployment of Asynchronous Robotic Sensors in Unknown Orthogonal Environments*

Eduardo Mesa Barrameda[1], Shantanu Das[2], and Nicola Santoro[3]

[1] Universidad de La Habana, Cuba
eduardomesa@matcom.uh.cu
[2] ETH Zurich, Switzerland
shantanu.das@inf.ethz.ch
[3] Carleton University, Canada
santoro@scs.carleton.ca

Abstract. We consider the problem of uniformly dispersing mobile robotic sensors in a simply connected orthogonal space of unknown shape. The mobile sensors are injected into the space from one or more entry points and rely only on sensed local information within a restricted radius. Unlike the existing solution, we allow the sensors to be asynchronous and show how, even in this case, the sensors can uniformly fill the unknown space, avoiding any collisions and without using any explicit communication, endowed with only $O(1)$ bits of persistent memory and $O(1)$ visibility radius. Our protocols are memory- and radius- optimal; in fact, we show that filling is impossible without persistent memory (even if visibility is unlimited); and that it is impossible with less visibility than that used by our algorithms (even if memory is unbounded).

1   Introduction

The Framework: An important problem for wireless sensor systems is the effective deployment of the sensors within the target space S. The deployment must usually satisfy some optimization criteria with respect to the space S (e.g., maximize coverage). In case of static sensors, they are usually deployed by external means, either carefully (e.g., manually installed) or randomly (e.g., dropped by an airplane); in the latter case, the distribution of the sensors may not satisfy the desired optimization criteria.

If the sensing entities are *mobile*, as in the case of mobile sensor networks, vehicular networks, and robotic sensor networks, they are potentially capable to position themselves in appropriate locations without the help of any central coordination or external control. However to achieve such a goal is a rather complex task, and designing localized algorithms for efficient and effective deployment of the mobile sensors is a challenging research issue.

We are interested in a specific instance of the problem, called the *Uniform Dispersal* (or *Filling*) problem, where the sensors have to completely fill an unknown

---

* Research partially supported by NSERC Canada.

S. Fekete (Ed.): ALGOSENSORS 2008, LNCS 5389, pp. 125–140, 2008.

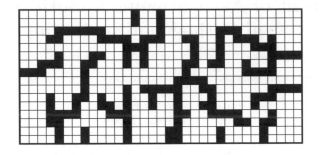

Fig. 1. An orthogonal region to be filled by the sensors

space S entering through one or more designated entry points called *doors*. In the process, the sensors must avoid colliding with each other, and must terminate (i.e., reach a quiescent state) within finite time. The space S is assumed to be simply connected (i.e., without holes), and orthogonal, i.e. polygonal with sides either parallel or perpendicular to one another (e.g., see Figure 1). Orthogonal spaces are interesting because they can be used to model indoor and urban environment.

We wish to study the problem from an algorithmic point of view, focussing on the minimum capabilities required by the sensors in order to effectively complete this task. We consider this problem within the context of robotic sensors networks: the mobile entities rely only on sensed local information within a restricted radius, called *visibility range*; when active they operate in a *sense-compute-move* cycle; and usually they have no explicit means of communication.

Existing Results: The problem of deployment of *mobile sensor networks* has been studied by several authors and continues to be the subject of extensive research; e.g., see [7,8,9,10,13,16,22]. Most of the work is focused on the *uniform self-deployment* problem; that is, how to achieve uniform deployment in S (usually assumed to be polygonal) starting from an initial random placement of the sensors in S. The *uniform dispersal* problem, studied here, has been previously investigated by Howard et al. [9]: sensors are deployed one-at-a-time into an unknown environment, and each sensor uses information gathered by previously deployed sensors to determine its deployment location.

The *robotic sensor networks*, studied in this paper, have been and continue to be the object of extensive investigations both from the control and the computing point of view (e.g., [1,3,4,5,6,12,14,20,21]; see [2,18] for recent surveys). A crucial difference between robotic sensor networks and traditional wireless sensor networks is in the determination of an entity's neighbours. In robotic sensor networks, the determination of one's neighbours is done by sensing capabilities (e.g., vision): any sensor in the sensing radius is detected even if inactive. On the other hand, in traditional wireless sensor networks, determination of the neighbours is achieved by radio communication; since an inactive sensor does not participate in any communication, the simple activity of determining one's neighbours, to be completed, requires the use of randomization or the presence of sophisticated synchronization and scheduling mechanisms (e.g., [15,17]). Both problems, *uniform self-deployment* and *uniform dispersal* have been studied for robotic sensor networks.

The *uniform self-deployment* problem for robotic sensor networks has been studied recently, and localized solution algorithms have been developed when the space S is a *line* (e.g., a rectilinear corridor) [3], and when it is a *ring* (e.g., the boundary of a convex region) [4]. The proposed solutions operate even if the sensors are very weak; indeed they are *anonymous* (i.e., indistinguishable), *oblivious* (i.e., without any recollection of computations and actions performed in the previous activity cycles), *asynchronous* (i.e., when the time between successive activity cycles is finite but unpredictable), and are *communication-free* (i.e., they use no explicit form of communication).

The *uniform dispersal* problem for robotic sensor networks, in which the sensors are injected one-at-a-time into the unknown environment S, has been introduced and investigated by Hsiang et al.[11]. Their results are based on an ingenious follow-the-leader technique where each sensor communicates with the one following it and instructions to move are communicated from predecessor to successor. The sensors are anonymous but they need some *persistent memory* to remember whether or not is a leader[1] and the direction of its movement. Since the algorithm uses only $O(1)$ bits of working memory in total, computationally the sensors can be just *finite-state machines*. In addition to requiring explicit communication, the solution of [11] makes the strong assumption that the sensors operate *synchronously*, which allows perfect coordination between the sensors.

This fact opens a series of interesting questions on the capabilities needed by the sensors to achieve uniform dispersal in orthogonal environments. In other words, how "weak" the sensors can be and still be able to uniformly disperse ? In particular: is persistent memory needed ? can the task be performed if the sensors are asynchronous ? is explicit communication really necessary ? In this paper we consider precisely these questions and provide some definite answers.

Our Results: We identify intrinsic limitations on the type of memory and the visibility radius of needed by asynchronous sensors to solve the uniform dispersal problem for any simply connected orthogonal space whose shape is a priori unknown. We then show that these results are tight; in fact we present protocols and prove that they solve the problem while meeting these limitations, and without using explicit communication.

We first consider the case of a *single door*. We show that oblivious sensors can *not* deterministically solve the problem, even if they have unlimited visibility. We then present (in section 4) an algorithm for solving the problem in the case of single door with asynchronous identical sensors having persistent working memory of only two bits and a visibility radius of just one unit.

For the case of *multiple doors*, we prove that asynchronous sensors can *not* solve the problem if the visibility radius is less than two units, even if they

---

[1] There is one leader for each door.

have unbounded memory. On the other hand, even with unbounded visibility and memory, the problem is still unsolvable if the sensors are identical. Thus, we assume that sensors entering the space from different doors have different colors (i.e. they are distinguishable). We prove that under this assumption, the problem can be solved with sensors having visibility radius two and constant (persistent) memory. The proof is constructive: we present the radius-optimal and memory optimal distributed algorithm for achieving this (in section 5).

Let us stress that, in our algorithms sensors use only constant memory; hence, they can be simple finite-state machines like in [11]. Unlike those in [11], our algorithms work for the asynchronous case; hence they are robust against occasional stalling of the sensors.

2   Model, Definitions and Properties

2.1   Sensors and Dispersal

The space to be filled by the sensors is a simply connected orthogonal region S that is partitioned into square cells each of size roughly equal to the area occupied by a sensor. Simply connected means that it is possible to reach any cell in the space from any other cell and there are no obstacle surrounded completely by cells belonging to the space.

The system is composed of simple entities, called sensors, having sensory and locomotion capabilities. The entities can turn and move in any direction. The sensory devices on the entity allows it to have a vision of its immediate surrounding; we assume the sensors to have restricted vision up to a fixed radius around it[2]. Even if two sensors see each-other, they do not have any explicit means of communicating with each-other. Each sensor functions according to an algorithm preprogrammed into it. The sensors have a $O(1)$ bits of working memory, and they have a local sense of orientation (i.e., each sensor has a consistent notion of "up-down" and "left-right");

If two sensors are in the same cell at the same time then there is a *collision*. The algorithm executed by the sensors must avoid collisions (e.g., to prevent damage to the sensor or its sensory equipment).

The sensors enter the space through special cells called doors [11]. A door is simply a cell in the space which always has a sensor in it. Whenever the sensor in the door moves to a neighboring cell, a new sensor appears instantaneously in the door. A sensor may not distinguish a door cell from an ordinary cell, using its sensory vision.

During each step taken by a sensor, the sensor first looks at its surrounding (up to its visibility radius) and then based on the rules of the algorithm, the sensor either chooses one of the neighboring cells to move to, or decides to remain stationary. Each step is atomic and during one step a sensor can only move to a neighboring cell. However, since the sensors are asynchronous, an arbitrary amount of time may lapse between two steps taken by a sensor.

---

[2] A visibility radius of one means that the robot sees all eight neighboring cells.

The problem to be solved is that of *uniform dispersal* (or *filling*) : within finite time, the entire space must be *filled*, i.e., every cell of the space is occupied by a sensor; furthermore the system configuration at that time must be *quiescent*, i.e., no sensor moves thereafter. The goal is to design a protocol P, the same for all sensors, that specifies which operations a sensor must perform whenever it is active, and that will always correctly within a finite time and without collisions lead the system to a quiescent configuration where the entire space is filled.

2.2   Space Representation and Properties

Let $A = A_y \times A_x$ be the smallest rectangular area containing the space S. We consider a partition of the area A into pixels $p_{i,j}$, $1 \leq i \leq r$, $1 \leq j \leq c$ where $r = A_y/q$, $c = A_x/q$ and q is the length of each cell in the space. Thus, some of the pixels (called valid pixels) correspond to the cells in the space while the other pixels represent obstacles. We represent the structure of the space S in the form a graph $G = (N, E)$ defined as follow:

- Each column of the space is partitioned into segments of consecutive valid pixels ended by an obstacle in both extremes and numbered from top to down.
- Each segment is a node of G. We denote by $l_j^k \in N$ the node corresponding to the $k - th$ segment of column j.
- We denote by $d\_p_j^k$ the bottom-most pixel of the segment l_j^k.
- There is an edge $(l_j^k, l_{j'}^{k'}) \in E$ if and only if:
    (a) $j = j' + 1$ or $j = j' - 1$ **and**
    (b) There is a pixel $p_{i,j'} \in l_{j'}^{k'}$ neighbor to $d\_p_j^k$ or there is a pixel $p_{i,j} \in l_j^k$ neighbor to $d\_p_{j'}^{k'}$.

The following propositions are easy to verify.

Proposition 1. *If l_j^k and $l_{j'}^{k'}$ are two distinct nodes of G then there is at most one edge between them.*

Proposition 2. *Two nodes l_j^k and $l_{j'}^{k'}$ have neighboring pixels if and only if there is an edge $(l_j^k, l_{j'}^{k'}) \in E$ between them.*

Proposition 3. *The graph G is connected.*

Proposition 4. *The graph G is acyclic.*

If there is an edge $(l_j^k, l_{j'}^{k'})$ such that the bottom-most pixel $d\_p_j^k = p_{i,j}$ of l_j^k is a neighbor of the pixel $p_{i,j'} \in, l_{j'}^{k'}$, we say that $p_{i,j}$ is the **entry point** from l_j^k to $l_{j'}^{k'}$ and $p_{i,j'}$ is the **entry point** from $l_{j'}^{k'}$ to l_j^k.

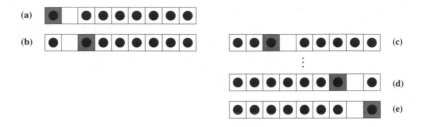

Fig. 2. (a) and (b) are indistinguishable configurations. (c) to (d) are allowable configurations; (e) is an unavoidable configuration.

3   Impossibility Results

We first show that the sensors must have some *persistent* memory of the past, for solving the filling problem successfully.

Theorem 1. *The filling problem can not be solved by oblivious sensors, even if they have unbounded visibility. This result holds even if there is only a single door.*

Proof. Consider the space consisting of a single line of $n = 2m+1$ pixels of which one of them is a door. By contradiction, let P be a correct filling protocol. Since the sensors have no memory of past, each step taken by a sensor depends only on the current configuration (i.e. which cells are filled and which are empty). We can represent each empty cell by 0 and each filled cell by 1; the door would be represented by D; however note that it is not distinguishable from a filled cell. A configuration can thus represented by the sequence $< d_1...d_n >$ of the the values of the cells left-to-right. If algorithm P is correct then the penultimate configuration (i.e., the final configuration before the space is completely filled), must have exactly one empty cell and this cell should be adjacent to the door. So, if the door is the leftmost cell then the only possible final configuration is $< D011....11111 >$. Notice that this is indistinguishable from the configuration $< 10D11....1111 >$ and the algorithm must make the same move in both cases. In the former situation, the leftmost robot (from the door) must move to the right, but the same move will leave the space unfilled in the latter scenario. So the configuration $< 10D11....1111 >$ must be avoided by the algorithm; this implies that the only correct penultimate configuration when the door is the third cell is $< 11D01....1111 >$. Extending this argument inductively, the only correct penultimate configuration when the door is the $2i + 1$-th cell $(0 \leq i < m)$, is the one where $d_{2i+1} = D$, $d_{2(i+1)} = 0$, and all other d_j's are 1. Hence the only correct penultimate configuration when the door is the $2(m-1)+1$-th cell, must be $< 11111....1D01 >$. Notice that this configuration is indistinguishable from $< 11111....110D >$ which thus must be avoided by the algorithm, However this is

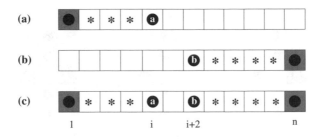

(a)

(b)

(c)

1   i   i+2   n

Fig. 3. A single line of cells with doors at one or both ends

the only possible penultimate configuration when the door is the rightmost cell. A contradiction.

Theorem 2. *In the case of multiple doors, it is impossible for asynchronous sensors to solve the filling problem avoiding collisions, if the visibility radius is less than two. The result holds even if the sensors have distinct visible identities and each sensor has an unbounded amount of memory.*

Proof. (Sketch) By contradiction, let P be a correct filling protocol for asynchronous sensors in simply-connected orthogonal spaces with multiple doors. Consider the space consisting of a single line of pixels, with a door at each end. Since the sensors at one door do not know of the existence of the other door, P must force the sensor initially at the left door (sensor 'a') to move towards the right end of the line (as in the case when there is no right door: figure 3(a)). Similarly, the protocol must force the sensor initially at the right door (sensor 'b') to move into the corridor towards the left end of the line (figure 3(b)). Thus, if both doors were present the sensors 'a' and 'b' must move towards each other. It is possible for an adversary to schedule the activations of the sensors (i.e., to choose the finite delays between successive activity cycles) so as to generate the situation where both 'a' and 'b' are about to enter the same empty cell that lies between them (see figure 3(c)). Since the visibility is less than two, they do not see each other but only the empty cell. By scheduling both sensors to move at the same time, the adversary generates a collision during the execution of P; this contradicts the assumption that P is correct and thus collision-free. Notice that even if the sensors remember a complete history of their past moves and have distinct visible IDs, the collision can not be avoided.

Theorem 3. *If sensors entering from distinct doors are indistinguishable from each other, then there is no collision-free solution to the filling problem for multiple doors. The result holds irrespective of the visibility range of the sensors.*

Proof. Consider the space S consisting of a single column of $n - 2$ cells ($n > 2$) and two extra cells adjacent to the column—one on the left and one on the right (see Figure 4). We consider three cases: (i) The cell on the left of the column

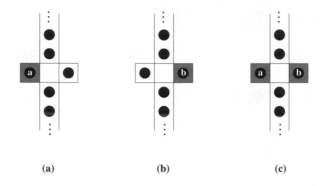

Fig. 4. (a) Single door on the left (b) Single door on the right (c) Indistinguishable sensors entering the column from two doors

is the only door, (ii) The cell on the right is the only door, and (iii) Both the cells to the left and the right of the column are doors. As before we consider the penultimate configuration reached by any correct filling algorithm for each of these cases, shown in Figure 4(a), (b) and (c) respectively. Notice that there is only one cell adjacent to the door in each case and thus this cell must be empty in the penultimate configuration. For cases (i) and (iii), let 'a' be the sensor currently at the door on the left. For cases (ii) and (iii), let 'b' be the sensor currently at the door on the right. It is possible for an adversary to schedule the activation of the sensors in such a way that both 'a' and 'b' become active for the first time only after reaching the penultimate configuration. Thus, neither sensor has any past history and any decision taken by sensor 'a' (or sensor 'b') would depend only on the current configuration. Notice that the current configuration in the three cases are indistinguishable from each other. So, sensor 'a' must take the same decision for case (i) and (iii). Similarly, sensor 'b' must take the same decision in case (ii) and (iii). If one of the sensors decides not to move to the empty cell, then the space remains unfilled in at least one of scenarios. On the other hand, if both decide to move, the adversary can force a collision for case (iii), by scheduling them to move at the same time.

In Section 5 we show that if the sensors have visibility radius at least two and sensors coming from distinct doors are distinguishable then there is a solution to the filling problem for any connected space with any number of doors.

4   Filling Algorithm: Single Door

In this section we consider the case when is only one door through which the sensors enter the space. We show that visibility radius of one and a constant amount of memory for each sensor, is sufficient in this case. Each sensor just needs one bit of memory, to remember its last location, so that it never backtracks. The idea of the algorithm (SINGLE) is to move the robots along the

---

Algorithm 1. SINGLE
___

Meta-Rule:

– A sensor never backtracks.

Rules: A sensor r in pixel $p_{i,j}$ executes the following rules:
  if ($p_{i+1,j}$ is empty) **then**
    r moves to $p_{i+1,j}$.
  else if ($p_{i-1,j}$ is empty) **then**
    r moves to $p_{i-1,j}$.
  else if (($p_{i,j-1}$ is empty) $\wedge$
       (($p_{i-1,j-1}$ is **obstacle**) $\vee$ ($p_{i-1,j}$ is **obstacle**))) **then**
    r moves to $p_{i,j-1}$.
  else if (($p_{i,j+1}$ is empty) $\wedge$
       (($p_{i-1,j+1}$ is **obstacle**) $\vee$ ($p_{i-1,j}$ is **obstacle**))) **then**
    r moves to $p_{i,j+1}$.
  else
    r does not move.
  end if

---

paths in G, starting from the node containing the door. Since the sensor can see the eight neighboring pixels, it can determine when it has reached an *entry point*.

Following the rules of algorithm SINGLE, any path of consecutive pixels in the space on which a sensor is allowed to travel is called a *valid path*. Notice that any valid path corresponds to some path in G.

Theorem 4. *Algorithm SINGLE solves the filling problem for any space of size n and a single door, without any collisions and using n sensors each having a constant amount of memory and a visibility radius of one.*

5 Filling Algorithm: Multiple Doors

If there are multiple doors, then we know that the sensors must have a visibility radius of at least two and they should not be indistinguishable. We assume sensors coming from different doors have different colors and each sensor has visibility radius of two. Our algorithm for this case, uses the following restriction on the movement of the sensors.

Meta-Rule. "A sensor may not move until its previous position is occupied".

The idea of our algorithm (called algorithm MULTIPLE) is sketched here. Sensors coming from different doors (i.e. sensors of different colors) follow distinct paths in G and these paths do not intersect. In other words, we ensure that the cells visited by sensors of color c_i are occupied by sensors of the same color (and never by sensors of any other color). Thus, a sensor before moving to a pixel $p_{i,j}$ needs to determine if this pixel was visited by sensors of another color. We show

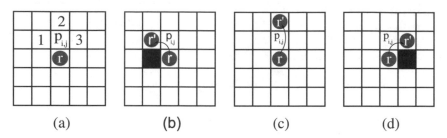

Fig. 5. Sensor r needs to check cells 1,2, and 3. The sensor (of the same color) that last visited $p_{i,j}$ must be in one of these locations.

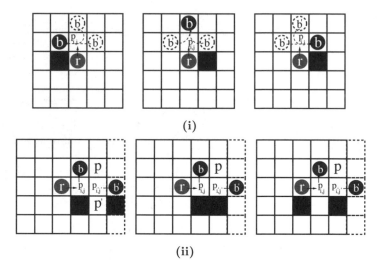

Fig. 6. (i) To determine if $p_{i,j}$ was visited by sensor b, sensor r should be able to see both b and its successor b'. (ii) If b' has moved out of the visibility range of r then there must be a sensor of the same color in cell P (or P').

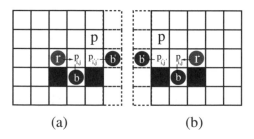

Fig. 7. If b' has moved out of the visibility range of r then there must be a sensor of the same color in the cell marked P

a sensor r of color c_r can always determine if the next pixel $p_{i,j}$ (in one of its valid paths) was visited by sensors of the same color (see Figure 5). In that case, the sensor r moves to the pixel $p_{i,j}$. Otherwise there may be two cases: (i) either pixel $p_{i,j}$ was visited by sensors of another color or (ii) pixel $p_{i,j}$ was never visited before. In the first case, sensor r does not move to pixel $p_{i,j}$ and searches for alternate paths. In the second case, sensor r needs to take a decision based on whether there are other sensors waiting to move into pixel $p_{i,j}$. In case there are two or more sensors in cells neighboring an unvisited pixel $p_{i,j}$, the sensors are assigned priorities[3] based whether they are coming from the left, right, top or bottom (in that order). The sensor with the highest priority (among those sensors for whom $p_{i,j}$ is a valid move) moves to pixel $p_{i,j}$. Notice that it may not be always possible for a sensor r to determine if its neighbor pixel $p_{i,j}$ is unvisited or has been visited by a sensor of another color (see Figure 6). If that is the case, then sensor r simply waits until the situation changes (so that it is able to take a decision). We shall show that such waiting never results in a deadlock, as the sensor with the highest priority to move is always able to decide, without waiting.

The formal description of algorithm MULTIPLE is given in Algorithm 2. The functions isValidUp(), isValidDown(), isValidLeft() and isValidRight() express how the sensor makes a decision if it can move or not to an specific neighboring pixel. We use the following notations in the algorithm:

- $r.Color$ is the color of the door sensor r came from.
- If a pixel $p_{i,j}$ is occupied by some sensor r, then $(p_{i,j}).Color = r.Color$.
  Otherwise $(p_{i,j}).Color = None$.

Algorithm 2. MULTIPLE

Meta-Rules:

- A sensor never backtracks.
- A sensor does not move if its previous position is empty.

Rules: A sensor r in a pixel $p_{i,j}$ of l_j^k executes the following:
if isValidUp(r) then
    r moves to $p_{i+1,j}$.
else if isValidDown(r) then
    r moves to $p_{i-1,j}$.
else if isValidLeft(r) then
    r moves to $p_{i,j-1}$.
else if isValidRight(r) then
    r moves to $p_{i,j+1}$.
else
    r does not move.
end if

Proposition 5. *The following properties hold during the execution of the algorithm MULTIPLE.*

---

[3] This is the only way to avoid collision as well as ensure progress of the algorithm.

Algorithm 3. Function IsValidDown($<sensor>$)

```
bool IsValidDown( r ){
    if p_{i-1,j} is occupied then
        return false
    else if ( (p_{i-2,j} is empty) ∧
              ( (p_{i-1,j-1} is empty) ∨ ((p_{i-2,j-1} is not obstacle) ∨
                                         (p_{i-2,j} is not obstacle   )) ) ∧
              ( (p_{i-1,j+1} is empty) ∨ ((p_{i,j+1} is not obstacle) ∨
                                         (p_{i-2,j} is not obstacle))) ) then
        return true
    else if ( ((p_{i-2,j}).Color = r.Color) ∨
              ( ((p_{i-2,j-1} is obstacle) ∨ (p_{i-2,j} is obstacle) ) ∧
                ((p_{i-1,j-1}).Color = r.Color)) ∨
              ( ((p_{i-2,j+1} is obstacle) ∨ (p_{i-2,j} is obstacle) ) ∧
                ((p_{i-1,j+1}).Color = r.Color)) ) then
        return true
    else
        return false
    end if
}
```

Algorithm 4. Function IsValidUp($<sensor>$)

```
bool IsValidUp( r ){
    if p_{i+1,j} is occupied then
        return false
    else if ( (p_{i+2,j} is empty) ∧
              ( (p_{i+1,j-1} is empty) ∨ (p_{i,j-1} is not obstacle) ) ∧
              ( (p_{i+1,j+1} is empty) ∨ (p_{i,j+1} is not obstacle) ) ) then
        return true
    else if ( ((p_{i+2,j}).Color = r.Color) ∨
              ((p_{i,j-1} is obstacle) ∧ ((p_{i+1,j-1}).Color = r.Color)) ∨
              ((p_{i,j+1} is obstacle) ∧ ((p_{i+1,j+1}).Color = r.Color)) ) then
        return true
    else if ( ((p_{i,j-1} is obstacle) ∧ (p_{i+1,j-1} is occupied)) ∨
              ((p_{i,j+1} is obstacle) ∧ (p_{i+1,j+1} is occupied)) ) then
        return false
    else if ( ( (p_{i,j-1} is obstacle) ∧
                (((p_{i+1,j-1}).Color = (p_{i+2,j}).Color) ∨
                 ((p_{i+2,j-1}).Color = (p_{i+2,j}).Color) ∨
                 ((p_{i+1,j-2}).Color = (p_{i+2,j}).Color))) ∨
              ( (p_{i,j+1} is obstacle) ∧
                (((p_{i+1,j+1}).Color = (p_{i+2,j}).Color) ∨
                 ((p_{i+2,j+1}).Color = (p_{i+2,j}).Color) ∨
                 ((p_{i+1,j+2}).Color = (p_{i+2,j}).Color))) ) then
        return false
    else
        return true
    end if
}
```

Algorithm 5. Function IsValidLeft($<sensor>$)

```
bool IsValidLeft( r ){
```
   if (($p_{i,j-1}$ is **occupied**) $\vee$
       (($p_{i-1,j-1}$ is not **obstacle**) $\wedge$ ($p_{i-1,j}$ is not **obstacle**))) **then**
     return false
   else if (($p_{i+1,j-1}$ is **empty**) $\wedge$ ($p_{i-1,j-1}$ is **empty**) $\wedge$
       (($p_{i,j-2}$ is **empty**) $\vee$
         (($p_{i-1,j-2}$ is not **obstacle**) $\wedge$ ($p_{i-1,j-1}$ is not **obstacle**)))) **then**
     return true
   else if ((($p_{i+1,j-1}).Color = r.Color$) $\vee$ (($p_{i-1,j-1}).Color = r.Color$) $\vee$
       ((($p_{i,j-2}).Color = r.Color$) $\wedge$
         (($p_{i-1,j-2}$ is **obstacle**) $\vee$ ($p_{i-1,j-1}$ is **obstacle**)))) **then**
     return true
   else if (($p_{i,j-2}$ is **occupied**) $\wedge$
       (($p_{i-1,j-2}$ is **obstacle**) $\vee$ ($p_{i-1,j-1}$ is **obstacle**)))) **then**
     return false
   else if (((($p_{i+1,j-1}).Color = (p_{i+1,j-2}).Color$) $\wedge$
       (($p_{i-1,j-2}$ is **obstacle**) $\vee$ ($p_{i-1,j-1}$ is **obstacle**))) $\vee$
       (($p_{i+1,j-1}).Color = (p_{i-1,j-2}).Color$) $\wedge$
       (($p_{i-1,j-1}$ is **obstacle**) $\vee$ ($p_{i-2,j-2}$ is **obstacle**) $\vee$
        ($p_{i-2,j-1}$ is **obstacle**))) $\vee$
       (($p_{i+1,j-1}).Color = (p_{i-1,j-1}).Color$) $\vee$
       (($p_{i+1,j-1}).Color = (p_{i-2,j-1}).Color$) $\vee$
       ((($p_{i-1,j-1}).Color = (p_{i+1,j-2}).Color$) $\wedge$
        (($p_{i-1,j-2}$ is **obstacle**) $\vee$ ($p_{i,j-2}$ is **obstacle**))) $\vee$
       (($p_{i-1,j-1}).Color = (p_{i+2,j-1}).Color$)) **then**
     return false
   else
     return true
   end if
```
}

---

(i)   *A sensor r can always determine if a neighboring pixel $p_{i,j}$ was visited by sensors from the same door.*

(ii)   *The sensor r having the highest priority to move into a pixel $p_{i,j}$ can always determine if the pixel $p_{i,j}$ is unvisited or not.*

(iii)   *Two sensors from different doors never visit the same pixel (i.e. no intersections).*

(iv)   *Two sensors are never in the same pixel at the same time (i.e. no collisions).*

**Proposition 6.** *The algorithm MULTIPLE terminates in a finite time.*

Based on the facts that there are no collisions and the sensors never re-visit the same pixel, we can prove this proposition in the same way as for the previous algorithm.

**Proposition 7.** *On termination of algorithm MULTIPLE, there are no empty pixels in the space.*
```

Algorithm 6. Function IsValidRight($<sensor>$)

---

bool IsValidRight(r){

   if (($p_{i,j+1}$ is **occupied**) $\vee$
      (($p_{i-1,j+1}$ is not **obstacle**) $\wedge$ ($p_{i-1,j}$ is not **obstacle**))) **then**
    return false
   else if (($p_{i+1,j+1}$ is **empty**) $\wedge$ ($p_{i-1,j+1}$ is **empty**) $\wedge$
      (($p_{i,j+2}$ is **empty**) $\vee$
        (($p_{i-1,j+2}$ is not **obstacle**) $\wedge$ ($p_{i-1,j+1}$ is not **obstacle**)))) **then**
    return true
   else if (((($p_{i+1,j+1}$).$Color$ = $r.Color$) $\vee$ (($p_{i-1,j+1}$).$Color$ = $r.Color$) $\vee$
      ((($p_{i,j+2}$).$Color$ = $r.Color$) $\wedge$
        (($p_{i-1,j+2}$ is **obstacle**) $\vee$ ($p_{i-1,j+1}$ is **obstacle**)))) **then**
    return true
   else if ((((($p_{i,j+2}$).$Color$ = ($p_{i+1,j+1}$).$Color$) $\vee$
        (($p_{i,j+2}$).$Color$ = ($p_{i+2,j+1}$).$Color$)) $\wedge$
      ($p_{i-1,j+2}$ is **obstacle**) $\vee$ ($p_{i-1,j+1}$ is **obstacle**))) $\vee$
      (((($p_{i,j+2}$).$Color$ = ($p_{i-1,j+1}$).$Color$) $\vee$
        (($p_{i,j+2}$).$Color$ = ($p_{i-2,j+1}$).$Color$)) $\wedge$
      ($p_{i-1,j+2}$ is **obstacle**)) $\vee$
      ((($p_{i+1,j+1}$).$Color$ = ($p_{i+1,j+2}$).$Color$) $\wedge$
      (($p_{i-1,j+2}$ is **obstacle**) $\vee$ ($p_{i-1,j+1}$ is **obstacle**))) $\vee$
      ((($p_{i+1,j+1}$).$Color$ = ($p_{i-1,j+2}$).$Color$) $\wedge$
      (($p_{i-1,j+1}$ is **obstacle**) $\vee$ ($p_{i-2,j+2}$ is **obstacle**) $\vee$
      ($p_{i-2,j+1}$ is **obstacle**))) $\vee$
      (($p_{i+1,j+1}$).$Color$ = ($p_{i-1,j+1}$).$Color$) $\vee$
      (($p_{i+1,j+1}$).$Color$ = ($p_{i-2,j+1}$).$Color$) $\vee$
      ((($p_{i-1,j+1}$).$Color$ = ($p_{i+1,j+2}$).$Color$) $\wedge$
      (($p_{i-1,j+2}$ is **obstacle**) $\vee$ ($p_{i,j+2}$ is **obstacle**))) $\vee$
      (($p_{i-1,j+1}$).$Color$ = ($p_{i+2,j+1}$).$Color$))) **then**
    return false
   else
    return true
   end if

}

Finally, we have the following result regarding the correctness of our algorithm.

Theorem 5. *The algorithm MULTIPLE completely fills any connected space, without collisions, even when the sensors enter from multiple doors (assuming they have distinct colors). The algorithm requires n sensors each having visibility radius two and a constant amount of memory.*

6 Conclusions and Open Problems

We have shown that, for uniform dispersal in simply connected orthogonal spaces, synchronicity and explicit communication are not necessary, while persistent memory is needed. More precisely, we have presented localized algorithms (one in the case of a single entry point, and one in the case of multiple entry points) that allow asynchronous mobile sensors to fill simply connected orthogonal spaces of unknown shape; the sensors do so without collisions and without any explicit direct communication, endowed with only $O(1)$ bits of persistent memory and $O(1)$ visibility radius. In both cases, the protocols are memory- and radius- optimal; in fact, we have shown that filling is impossible without persistent memory (even if visibility is unlimited); it is also impossible with less visibility than that used by our algorithms (even if memory is unbounded).

There are many interesting research problem still open. For example, orthogonal spaces that are not simply connected (i.e., containing holes) can be filled by synchronous sensors [11], but asynchronous solutions are not yet available. The study of the filling problem in more general classes of spaces is still open both in the synchronous and asynchronous settings. Another interesting direction for future research is to study the impact of having communication capabilities.

References

1. Agmon, N., Peleg, D.: Fault-tolerant gathering algorithms for autonomous mobile robots. SIAM J. on Computing 36, 56–82 (2006)
2. Bullo, F., Cortes, J., Martinez, S.: Distributed algorithms for robotic networks. In: Meyers, R. (ed.) Encyclopedia of Complexity and Systems Science. Springer, Heidelberg (to appear, 2008)
3. Cohen, R., Peleg, D.: Local algorithms for autonomous robot systems. In: Proc. 13th Colloquium on Structural Information and Communication Complexity, pp. 29–43 (2006)
4. Flocchini, P., Prencipe, G., Santoro, N.: Self-deployment of mobile sensors on a ring. Theoretical Computer Science 402(1), 67–80 (2008)
5. Flocchini, P., Prencipe, G., Santoro, N., Widmayer, P.: Gathering of asynchronous mobile robots with limited visibility. Theoretical Computer Science 337, 147 168 (2005)
6. Ganguli, A., Cortes, J., Bullo, F.: Visibility-based multi-agent deployment in orthogonal environments. In: Proceedings American Control Conference, pp. 3426–3431 (2007)
7. Heo, N., Varshney, P.K.: A distributed self spreading algorithm for mobile wireless sensor networks. In: Proceedings IEEE Wireless Communication and Networking Conference, vol. 3, pp. 1597–1602 (2003)
8. Heo, N., Varshney, P.K.: Energy-efficient deployment of intelligent mobile sensor networks. IEEE Transactions on Systems, Man, and CyberNetics - Part A 35(1), 78–92 (2005)
9. Howard, A., Mataric, M.J., Sukahatme, G.S.: An incremental self-deployment algorithm for mobile sensor networks. IEEE Transactions on Robotics and Automation 13(2), 113–126 (2002)
10. Howard, A., Mataric, M.J., Sukhatme, G.S.: Mobile sensor network deployment using potential fields: A distributed, scalable solution to the area coverage problem. In: Proceedings 6th International Symposium on Distributed Autonomous Robotics Systems (DARS 2002), pp. 299–308 (2002)
11. Hsiang, T.R., Arkin, E., Bender, M.A., Fekete, S., Mitchell, J.: Algorithms for rapidly dispersing robot swarms in unknown environment. In: Proc. 5th Workshop on Algorithmic Foundations of Robotics (WAFR), pp. 77–94 (2002)

12. Lee, J., Venkatesh, S., Kumar, M.: Formation of a geometric pattern with a mobile wireless sensor network. Journal of Robotic Systems 21(10), 517–530 (2004)
13. Loo, L., Lin, E., Kam, M., Varshney, P.: Cooperative multi-agent constellation formation under sensing and communication constraints. Cooperative Control and Optimization, 143–170 (2002)
14. Martinson, E., Payton, D.: Lattice formation in mobile autonomous sensor arrays. In: Proc. International Workshop on Swarm Robotics (SAB 2004), pp. 98–111 (2004)
15. Nikoletseas, S.E.: Models and algorithms for wireless sensor networks. In: Proc. 32nd Conference on Current Trends in Theory and Practice of Computer Science, pp. 64–83 (2006)
16. Poduri, S., Sukhatme, G.S.: Constrained coverage for mobile sensor networks. In: Proc. IEEE Int. Conference on Robotic and Automation, pp. 165–173 (2004)
17. Powell, O., Leone, P., Rolim, J.: Energy optimal data propagation in wireless sensor networks. Journal of Parallel and Distributed Computing 67(3), 302–317 (2007)
18. Prencipe, G., Santoro, N.: Distributed algorithms for mobile robots. In: Proc. 5th IFIP International Conference on Theoretical Computer Science (TCS 2006) (2006)
19. Reif, J.H., Wang, H.: Social potential fields: A distributed behavioral control for autonomous robots. Robotics and Autonomous Systems 27(3), 171–194 (1999)
20. Susca, S., Martinez, S., Bullo, F.: Monitoring enviromental boundaries with a robotic sensor network. IEEE Transactions on Control Systems Technology 16(2), 288–296 (2008)
21. Suzuki, I., Yamashita, M.: Distributed anonymous mobile robots: Formation of geometric patterns. SIAM J. Comput. 28(4), 1347–1363 (1999)
22. Wang, G., Cao, G., La Porta, T.: Movement-assisted sensor deployment. In: Proc. IEEE INFOCOM, vol. 4, pp. 2469–2479 (2004)

Optimal Backlog in the Plane

Valentin Polishchuk and Jukka Suomela

Helsinki Institute for Information Technology HIIT
Helsinki University of Technology and University of Helsinki
P.O. Box 68, FI-00014 University of Helsinki, Finland
valentin.polishchuk@cs.helsinki.fi, jukka.suomela@cs.helsinki.fi

Abstract. Suppose that a cup is installed at every point of a planar set P, and that somebody pours water into the cups. The total rate at which the water flows into the cups is 1. A player moves in the plane with unit speed, emptying the cups. At any time, the player sees how much water there is in every cup. The player has no information on how the water will be poured into the cups in the future; in particular, the pouring may depend on the player's motion. The *backlog* of the player is the maximum amount of water in any cup at any time, and the player's objective is to minimise the backlog. Let D be the diameter of P. If the water is poured at the rate of $1/2$ into the cups at the ends of a diameter, the backlog is $\Omega(D)$. We show that there is a strategy for the player that guarantees the backlog of $O(D)$, matching the lower bound up to a multiplicative constant. Note that our guarantee is independent of the number of the cups.

1 Introduction

Consider a wireless sensor network where each sensor node stores data locally on a flash memory card [1,2,3,4,5]. To conserve energy, radio communication is only used for control information; for example, each network node periodically reports the amount of data stored on the memory card. We have one maintenance man who visits the sensors and physically gathers the information stored on the memory cards. The maximum amount of data produced at a sensor between consecutive visits – the *backlog* – determines how large memory cards are needed. In this work we study the problem of designing a route for the maintenance man to minimise the backlog.

If each device produces data at the same constant rate, then we have an offline optimisation problem which is exactly the travelling salesman problem: find a minimum-length tour that visits each device.

However, in the general case, the data rate may vary from time to time and from device to device; for example, a motion-tracking application produces data only when there is a monitored entity in the immediate neighbourhood of the sensor. We arrive at an *online* problem: guide the maintenance man so that the maximum backlog is minimised, using only the information on the current backlog.

S. Fekete (Ed.): ALGOSENSORS 2008, LNCS 5389, pp. 141–150, 2008.

We only assume that we know the *total* volume of the data that is gathered on the memory cards in one time unit; we assume nothing about the distribution of the data. For example, our network could be tracking a constant number of moving objects, and each of the objects produces data at a constant rate at its nearest sensor; we assume nothing about how the objects move. At first glance assuming so little may seem overly cautious, but as we shall see, we can nevertheless obtain strong positive results.

Our primary interest is in the scalability of the system. Trivially, doubling the total rate of the data means that the backlog may double as well. We focus on the effect of increasing the number of sensor devices and increasing the physical area within which the devices are installed.

Surprisingly, it turns out that if the diameter of the physical area is fixed and the total rate at which the data accumulates at the devices is fixed, we can guarantee a certain amount of backlog *regardless of the number* or *placement* of the sensors. The adversary can install any number of sensors in arbitrary locations, and the adversary can also change the distribution of the data depending on the activities of our maintenance man; nevertheless, we can keep the backlog below a certain level.

As a direct consequence, the backlog increases only linearly in the diameter of the physical area. If the diameter is doubled, we only need to double the speed of our maintenance man – or double the size of the memory cards.

1.1 Problem Formulation and Contribution

Formally, the *minimum backlog game* is defined as follows. Because of historical precedent [6,7], we use cups and water instead of sensor devices and data.

Definition 1 (Minimum backlog game). *Let $P \subset \mathbb{R}^2$ be a finite planar set. There is a cup on each point in P. The adversary pours water into the cups P; the total rate at which the water is poured into the cups is 1. The player moves in the plane with unit speed, starting at an arbitrary point. Whenever the player visits a cup, the cup is emptied. The goal of the player is to keep the maximum level of water in any cup at any time as low as possible.*

Let D be the diameter of P. Clearly, the adversary can ensure that some cup at some time will contain $\Omega(D)$ water: the adversary may just take two cups that define the diameter of P, and pour water with the rate $1/2$ into each of them. In this paper, we prove that the player can match this, up to a multiplicative constant.

Theorem 1. *The player has a strategy which guarantees that no cup ever contains more than $O(D)$ units of water.*

We emphasise that the result does not depend on the number of cups in the set, but just on its diameter.

1.2 Comparison with Previous Work

In the *discrete* version of the game [7], the cups are installed at the nodes of a graph. The game proceeds in *time steps*; in each step the adversary pours a total of 1 water, while the player moves from a node to an adjacent node. The competitive ratio of any algorithm has a lower bound of $\Omega(\Delta)$, where Δ is the diameter of the graph [7]. This motivated determining a strategy for the player that guarantees a uniform upper bound on the backlog.

In an arbitrary graph on n nodes, the deamortization analysis of Dietz and Sleator [8] leads to a strategy that guarantees $O(\Delta \log n)$ backlog. In certain graphs (stars) the adversary can guarantee a matching $\Omega(\Delta \log n)$ lower bound. The main contribution of Bender et al. [7] was an algorithm achieving a backlog of $O(n\sqrt{\log \log n})$ in the case when the graph is an n-by-n grid. It was sketched how to extend the regular-grid approach to the game on a general planar set P to guarantee a backlog of $O(D\sqrt{\log \log|P|})$, where D is the diameter of P.

Our strategy guarantees a backlog of $O(D)$, independent of the number of cups. The strategy consists of a set of coroutines, each responsible for clearing water of certain age. Similarly to prior work [7], we compare the performance of the coroutines to that of a player in an "imaginary" game. We give bounds on the performance of each of the coroutines, combining them into the bound for the topmost algorithm.

2 Preliminaries

We define the concept of a (τ, k)-game, forming the basis of our analysis. We give a preliminary lemma about the performance of the player in the game; the lemma is a direct extension of a result of Dietz and Sleator [8]. We also recall a result of Few [9] stating that for any planar set Q there exists a cycle through Q of length $O(\mathrm{diam}(Q)\sqrt{|Q|})$.

2.1 The (τ, k)-Game

For each $\tau \in \mathbb{R}$, $k \in \mathbb{N}$ we define the (τ, k)-game as follows.

Definition 2. *There is a set of cups, initially empty, not located in any particular metric space. At each time step the following takes place, in this order:*

1. *The adversary pours a total of τ units of water into the cups. The adversary is free to distribute the water in any way he likes.*
2. *The player empties k fullest cups.*

The game is discrete – the player and the adversary take turns making moves during discrete time steps. The next lemma bounds the amount of water in any cup after few steps of the game.

Lemma 1. *The water level in any cup after r complete time steps of the (τ, k)-game is at most $H_r\tau/k$, where H_r is the rth harmonic number.*

Proof. We follow the analysis of Dietz and Sleator [8, Theorem 5]. Consider the water levels in the cups after the time step j. Let $X_j^{(i)}$ be the amount of water in the cup that is ith fullest, and let

$$S_j = \sum_{i=1}^{(r-j)k+1} X_j^{(i)}$$

be the total amount of water in $(r - j)k + 1$ fullest cups at that time. Initially, $j = 0$, $X_0^{(i)} = 0$, and $S_0 = 0$.

Let us consider what happens during the time step $j \in \{1, 2, \ldots, r\}$. The adversary pours τ units of water; the total amount of water in $(r - j + 1)k + 1$ fullest cups is therefore at most $S_{j-1} + \tau$ after the adversary's move and before the player's move (the worst case being that the adversary pours all water into $(r - j + 1)k + 1$ fullest cups).

Then the player empties k cups. The k fullest cups contained at least a fraction $k/((r - j + 1)k + 1)$ of all water in the $(r - j + 1)k + 1$ fullest cups; the remaining $(r - j)k + 1$ cups are now the fullest. We obtain the inequality

$$S_j \leq \left(1 - \frac{k}{(r - j + 1)k + 1}\right)(\tau + S_{j-1})$$

or

$$\frac{S_j}{(r - j)k + 1} \leq \frac{\tau}{(r - j + 1)k + 1} + \frac{S_{j-1}}{(r - j + 1)k + 1}.$$

Therefore the fullest cup after time step r has the water level at most

$$X_r^{(1)} = \frac{S_r}{k(r - r) + 1}$$

$$\leq \frac{\tau}{1k + 1} + \frac{\tau}{2k + 1} + \ldots + \frac{\tau}{rk + 1} + \frac{S_0}{rk + 1}$$

$$\leq \frac{\tau}{k}\left(\frac{1}{1} + \frac{1}{2} + \ldots + \frac{1}{r}\right). \qquad \square$$

2.2   Few's Lemma

Our strategy for the minimum backlog game invokes a number of coroutines at different moments of time. The following result by Few [9] provides the basis of the proof that the execution of the coroutines can indeed be scheduled as we define.

Lemma 2. *[9, Theorem 1] Given n points in a unit square, there is a path through the n points of length not exceeding $\sqrt{2n} + 1.75$.*

We make use of the following corollary.

Corollary 1. *Let S be a $D \times D$ square. Let $i \in \{0, 1, \ldots\}$. Let $Q \subset S$ be a planar point set with $|Q| = 25^i$ and $\mathrm{diam}(Q) = D$. For any point $p \in S$ there exists a closed tour of length at most $5^{i+1}D$ that starts at p, visits all points in Q, and returns to p.*

Proof. If $i > 0$, by Lemma 2, there is tour of length at most

$$\left(\sqrt{2(25^i + 1)} + 1.75 + \sqrt{2}\right)D \leq 5^{i+1}D$$

that starts at p, visits all points in Q, and returns to p. If $i = 0$, there is a tour of length $2\sqrt{2}D \leq 5D$ through p and $|Q| = 1$ points. $\qquad \square$

3   The Strategy

The player's strategy is composed from a number of coroutines, which we label with $i \in \{0, 1, \dots\}$. The coroutine i is invoked at times $(10L + \ell)\tau_i$ for each $L \in \{0, 1, \dots\}$ and $\ell \in \{1, 2, \dots, 10\}$. Whenever a lower-numbered coroutine is invoked, higher-numbered coroutines are suspended until the lower-numbered coroutine returns.

We choose the values τ_i as follows. For $i \in \{0, 1, 2, \dots\}$, let

$$k_i = 25^i,$$
$$\tau_i = (2/5)^i \cdot 10Dk_i = 10^i \cdot 10D.$$

For $L \in \{0, 1, \dots\}$ and $\ell \in \{1, 2, \dots, 10\}$, define (i, L, ℓ)-*water* to be the water that was poured during the time interval $[10L\tau_i, (10L + \ell)\tau_i]$.

3.1   The Coroutine i

The coroutine i performs the following tasks when invoked at time $(10L + \ell)\tau_i$:

1. Determine which k_i cups to empty. The coroutine chooses to empty k_i cups with the largest amount of (i, L, ℓ)-water.
2. Choose a cycle of length at most $\tau_i/2^{i+1}$ which visits the k_i cups and returns back to the original position. This is possible by Corollary 1 because $5^{i+1}D = \tau_i/2^{i+1}$.
3. Guide the player through the chosen cycle.
4. Return.

Observe that when a coroutine returns, the player is back in the location from which the cycle started. Therefore invocations of lower-numbered coroutines do not interfere with any higher-number coroutines which are currently suspended; they just delay the completion of the higher-numbered coroutines.

The completion is not delayed for too long. Indeed, consider a time period $[j\tau_i, (j+1)\tau_i]$ between consecutive invocations of the coroutine i. For $h \in \{0, 1, \dots, i\}$, the coroutine h is invoked 10^{i-h} times during the period. The cycles of the coroutine h have total length at most

$$10^{i-h}\tau_h/2^{h+1} = \tau_i/2^{h+1}.$$

In grand total, all coroutines $0, 1, \dots, i$ invoked during the time period take time

$$\sum_{h=0}^{i} \tau_i/2^{h+1} < \tau_i.$$

Therefore all coroutines invoked during the time period are able to complete within it. This proves that the execution of the coroutines can be scheduled as described.

4   Analysis

We now analyse the backlog under the above strategy. For any points in time $0 \le t_1 \le t_2 \le t_3$, we write $W([t_1, t_2], t_3)$ for the maximum per-cup amount of water that was poured during the time interval $[t_1, t_2]$ and is still in the cups at time t_3.

We need to show that $W([0, t], t)$ is bounded by a constant that does not depend on t. To this end, we first bound the amount of (i, L, ℓ)-water present in the cups at the time $(10L + \ell + 1)\tau_i$. Then we bound the maximum per-cup amount of water at an arbitrary moment of time by decomposing the water into (i, L, ℓ)-waters and a small remainder.

We make use of the following simple properties of $W([\cdot, \cdot], \cdot)$. Consider any four points in time $0 \le t_1 \le t_2 \le t_3 \le t_4$. First, the backlog for *old* water is nonincreasing: $W([t_1, t_2], t_4) \le W([t_1, t_2], t_3)$. Second, we can decompose the backlog into smaller parts: $W([t_1, t_3], t_4) \le W([t_1, t_2], t_4) + W([t_2, t_3], t_4)$.

4.1 (i, L, ℓ)-Water at Time $(10L + \ell + 1)\tau_i$

Let $i \in \{0, 1, \dots\}$, $L \in \{0, 1, \dots\}$, and $\ell \in \{1, 2, \dots, 10\}$. Consider (i, L, ℓ)-water and the activities of the coroutine i when it was invoked at the times $(10L + 1)\tau_i$, $(10L + 2)\tau_i, \dots, (10L + \ell)\tau_i$.

The crucial observation is the following. By the time $(10L + \ell + 1)\tau_i$, the coroutine i has, in essence, played a (τ_i, k_i)-game for ℓ rounds with (i, L, ℓ)-water. The difference is that the coroutine cannot empty the cups immediately after the adversary's move; instead, the cup-emptying takes place during the adversary's next move. Temporarily, some cups may be fuller than in the (τ_i, k_i)-game. However, once we wait for τ_i time units to let the coroutine i complete its clean-up tour, the maximum level of the water that has arrived before the beginning of the clean-up tour is at most that in the (τ_i, k_i)-game.

The fact that the emptying of the cups is delayed can only hurt the adversary, as during the clean-up tour the player may also accidentally undo some of the cup-filling that the adversary has performed on his turn. The same applies to the intervening lower-numbered coroutines.

Therefore, by Lemma 1, we have

$$W\big([10L\tau_i, (10L + \ell)\tau_i], (10L + \ell + 1)\tau_i\big) \le H_\ell \cdot \tau_i/k_i < 3\tau_i/k_i. \qquad (1)$$

4.2   Decomposing Arbitrary Time t

Let t be an arbitrary instant of time. We can write t as $t = T\tau_0 + \epsilon$ for a nonnegative integer T and some remainder $0 \le \epsilon < \tau_0$. Furthermore, we can represent the integer T as

$$T = \ell_0 + 10\ell_1 + \dots + 10^N \ell_N$$

for some integers $N \in \{0, 1, \dots\}$ and $\ell_i \in \{1, 2, \dots, 10\}$. Since the range is $1 \le \ell_i \le 10$, not $0 \le \ell_i \le 9$, this is not quite the usual decimal representation; we chose this range for ℓ_i to make sure that ℓ_i is never equal to 0.

We also need partial sums

$$L_i = \ell_{i+1} + 10\ell_{i+2} + \cdots + 10^{N-i-1}\ell_N.$$

Put otherwise, for each $i \in \{0, 1, \ldots, N\}$ we have

$$T = \ell_0 + 10\ell_1 + \cdots + 10^i \ell_i + 10^{i+1} L_i$$

and therefore

$$\begin{aligned} t &= \epsilon + \ell_0 \tau_0 + \ell_1 \tau_1 + \cdots + \ell_N \tau_N \\ &= \epsilon + \ell_0 \tau_0 + \ell_1 \tau_1 + \cdots + \ell_i \tau_i + 10 L_i \tau_i. \end{aligned}$$

We partition the time from 0 to $t - \epsilon$ into long and short periods. The *long period* $i \in \{0, 1, \ldots, N\}$ is of the form $\left[10 L_i \tau_i, (10 L_i + \ell_i - 1)\tau_i \right]$ and the *short period* i is of the form $\left[(10 L_i + \ell_i - 1)\tau_i, (10 L_i + \ell_i)\tau_i \right]$. See Fig. 1 for an illustration. Short periods are always nonempty, but the long period i is empty if $\ell_i = 1$.

4.3   Any Water at Arbitrary Time t

Now we make use of the decomposition of an arbitrary time interval $[0, t]$ defined in the previous section: we have long periods $i \in \{0, 1, \ldots, N\}$, short periods $i \in \{0, 1, \ldots, N\}$, and the remainder $[t - \epsilon, t]$.

Consider the long period i. We bound the backlog from this period by considering the point in time $(10 L_i + \ell_i)\tau_i \leq t$. If the period is nonempty, that is, $\ell_i > 1$, then we have by (1)

$$\begin{aligned} &W\left(\left[10 L_i \tau_i, (10 L_i + \ell_i - 1)\tau_i \right], t \right) \\ &\leq W\left(\left[10 L_i \tau_i, (10 L_i + \ell_i - 1)\tau_i \right], (10 L_i + \ell_i)\tau_i \right) < 3\tau_i / k_i. \end{aligned} \tag{2}$$

Naturally, if the period is empty, then (2) holds as well.

Consider a short period i for $i > 0$. It will be more convenient to write the short period in the form $\left[10 L_i' \tau_{i-1}, (10 L_i' + 10)\tau_{i-1} \right]$ where $L_i' = 10 L_i + \ell_i - 1$ is a nonnegative integer (this is illustrated in Fig. 1 for $i = 3$). We bound the backlog from this period by considering the point in time $(10 L_i' + 11)\tau_{i-1} \leq t$. By (1) we have

$$\begin{aligned} &W\left(\left[(10 L_i + \ell_i - 1)\tau_i, (10 L_i + \ell_i)\tau_i \right], t \right) \\ &= W\left(\left[10 L_i' \tau_{i-1}, (10 L_i' + 10)\tau_{i-1} \right], t \right) \\ &\leq W\left(\left[10 L_i' \tau_{i-1}, (10 L_i' + 10)\tau_{i-1} \right], (10 L_i + 11)\tau_{i-1} \right) < 3\tau_{i-1} / k_{i-1}. \end{aligned} \tag{3}$$

Next consider the short period $i = 0$. We have the trivial bound τ_0 for the water that arrived during the period. Therefore

$$W\left(\left[(10 L_0 + \ell_0 - 1)\tau_0, (10 L_0 + \ell_0)\tau_0 \right], t \right) \leq \tau_0. \tag{4}$$

Finally, we have the time segment from $t - \epsilon$ to t. Again, we have the trivial bound ϵ for the water that arrived during the time segment. Therefore

$$W\left(\left[t - \epsilon, t \right], t \right) \leq \epsilon < \tau_0. \tag{5}$$

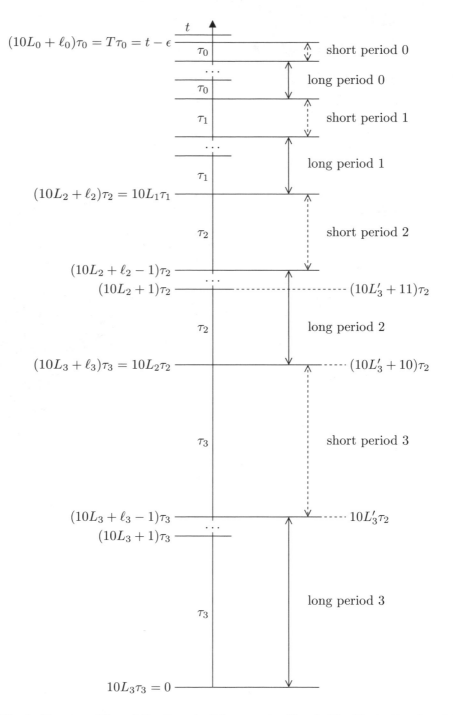

Fig. 1. Decomposition of the time; in this example, $N = 3$. The illustration is not in scale; actually $\tau_i = 10\tau_{i-1}$.

Now we can obtain an upper bound for the backlog at time t. Summing up (2), (3), (4), and (5), we have the maximum backlog

$$W([0, t], t) \leq \sum_{i=0}^{N} W\left(\left[10L_i\tau_i, (10L_i + \ell_i - 1)\tau_i\right], t\right)$$

$$+ \sum_{i=0}^{N} W\left(\left[(10L_i + \ell_i - 1)\tau_i, (10L_i + \ell_i)\tau_i\right], t\right)$$

$$+ W([t - \epsilon, t], t)$$

$$\leq \sum_{i=0}^{N} 3\tau_i/k_i + \sum_{i=1}^{N} 3\tau_{i-1}/k_{i-1} + 2\tau_0$$

$$\leq \sum_{i=0}^{\infty} 6\tau_i/k_i + 2\tau_0$$

$$= 60D \sum_{i=0}^{\infty} (2/5)^i + 20D = 120D = O(D).$$

5   Conclusions

We have shown that the player in the minimum backlog game has a strategy where the backlog does not depend on the number of cups, but only on the diameter of the cups set. This implies that the backlog scales linearly with the diameter of the area.

An interesting open question is the scalability in the *number of players*. If we have four players instead of one, we can divide the area into four parts and assign each player into one of the parts; this effectively halves the diameter and thus halves the backlog. It remains to be investigated whether we can exploit multiple players in a more efficient manner.

Acknowledgements

We thank Esther Arkin, Michael Bender, Sándor Fekete, Alexander Kröller, Vincenzo Liberatore, and Joseph Mitchell for discussions. This research was supported in part by the Academy of Finland, Grants 116547 and 118653 (ALGO-DAN), and by Helsinki Graduate School in Computer Science and Engineering (Hecse).

References

1. Somasundara, A.A., Ramamoorthy, A., Srivastava, M.B.: Mobile element scheduling for efficient data collection in wireless sensor networks with dynamic deadlines. In: Proc. 25th IEEE International Real-Time Systems Symposium (RTSS), Lisbon, Portugal, pp. 296–305. IEEE Computer Society Press, Los Alamitos (2004)

2. Jea, D., Somasundara, A., Srivastava, M.: Multiple controlled mobile elements (data mules) for data collection in sensor networks. In: Prasanna, V.K., Iyengar, S.S., Spirakis, P.G., Welsh, M. (eds.) DCOSS 2005. LNCS, vol. 3560, pp. 244–257. Springer, Heidelberg (2005)
3. Gu, Y., Bozdağ, D., Brewer, R.W., Ekici, E.: Data harvesting with mobile elements in wireless sensor networks. Computer Networks 50(17), 3449–3465 (2006)
4. Mathur, G., Desnoyers, P., Ganesan, D., Shenoy, P.: Ultra-low power data storage for sensor networks. In: Proc. 5th International Conference on Information Processing in Sensor Networks (IPSN), Nashville, TN, USA, pp. 374–381. ACM Press, New York (2006)
5. Diao, Y., Ganesan, D., Mathur, G., Shenoy, P.: Rethinking data management for storage-centric sensor networks. In: Proc. 3rd Biennial Conference on Innovative Data Systems Research (CIDR), Asilomar, CA, USA, pp. 22–32 (January 2007)
6. Adler, M., Berenbrink, P., Friedetzky, T., Goldberg, L.A., Goldberg, P., Paterson, M.: A proportionate fair scheduling rule with good worst-case performance. In: Proc. 15th Annual ACM Symposium on Parallel Algorithms and Architectures (SPAA), San Diego, CA, USA, pp. 101–108. ACM Press, New York (2003)
7. Bender, M.A., Fekete, S.P., Kröller, A., Liberatore, V., Mitchell, J.S.B., Polishchuk, V., Suomela, J.: The minimum-backlog problem. In: 2nd International Conference on Mathematical Aspects of Computer and Information Sciences (MACIS), Paris, France (December 2007)
8. Dietz, P.F., Sleator, D.D.: Two algorithms for maintaining order in a list. In: Proc. 19th Annual ACM Symposium on Theory of Computing (STOC), pp. 365–372. ACM Press, New York (1987)
9. Few, L.: The shortest path and the shortest road through n points. Mathematika 2, 141–144 (1955)

Author Index

Printing: Mercedes-Druck, Berlin
Binding: Stein+Lehmann, Berlin